This student workbook is intended to reinforce your learning of the content on the GCSE Physics specification (J635), from the OCR Twenty First Century Science Suite.

The workbook is cross-referenced to the revision and classroom companion *OCR Twenty First Century GCSE Physics Revision Plus* published by Letts and Lonsdale.

The pages for each unit are divided into topics to correspond with the specification and revision guide, and provide a clear, manageable structure to your revision. They focus on the material that is externally assessed (i.e. tested under exam conditions). They do not cover the practical data analysis and case study or the practical investigation, which are marked by your teacher.

You will have to sit three exams in total, including an Ideas in Context paper, which will test your ability to use and apply your scientific knowledge, for example, to understand and evaluate information about a current social-science issue. The questions on pages 75–76 of this workbook have been specially designed to allow you to practise these skills.

This workbook is suitable for use by Foundation and Higher Tier students.

> **HT** Any material that is limited to Higher Tier students appears inside a grey tinted box, clearly labelled with the symbol **HT**.

A Note to Teachers

The pages in this workbook can be used as…

- classwork sheets – students can use the revision guide to answer the questions
- harder classwork sheets – pupils study the topic and then answer the questions without using the revision guide
- easy-to-mark homework sheets – to test pupils' understanding and reinforce their learning
- the basis for learning homework tasks, which are then tested in subsequent lessons
- test materials for topics or entire units
- a structured revision programme prior to the objective tests / written exams.

Answers to this workbook are available to order.

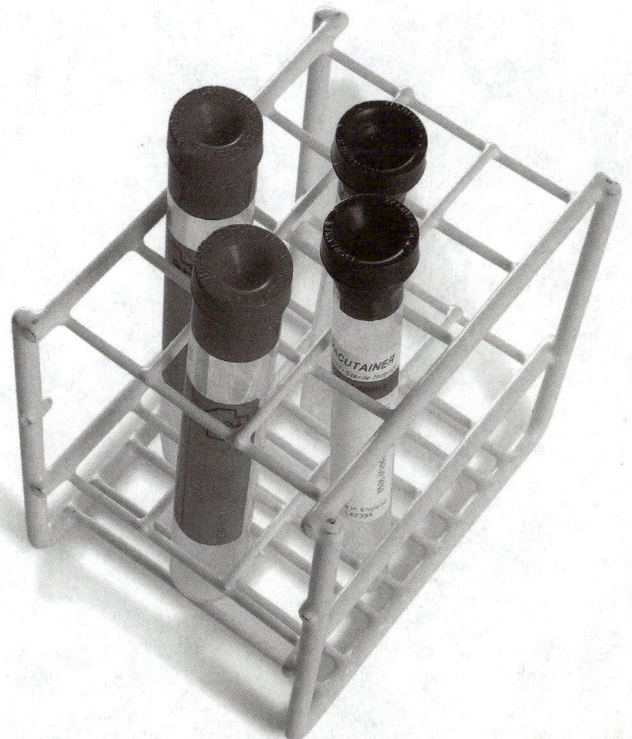

Contents

Contents

The Earth in the Universe

1 a) People used to believe that the Earth was only 6000 years old. Suggest two reasons for this.

i) _Scientists believed in this theory_

ii) _There was no way of testing or any evidence_

b) Evidence from rocks proves that the Earth is much older than 6000 years. Use the clues, about how rocks provide this evidence, to complete the crossword below.

Across

3. Radioactive _Dating_ is used to find the age of rocks (6)
4. When a huge force bends rocks into new shapes this is called _folding_ (7)
5. Breaks up big rocks and wears away the Earth's surface (7)

Down

1. Scientists now believe the Earth to be 4500 _million_ years old (7)
2. Earth does not have many of these but the Moon has lots of them (7)
4. These get trapped in sediment; most are millions of years old (7)

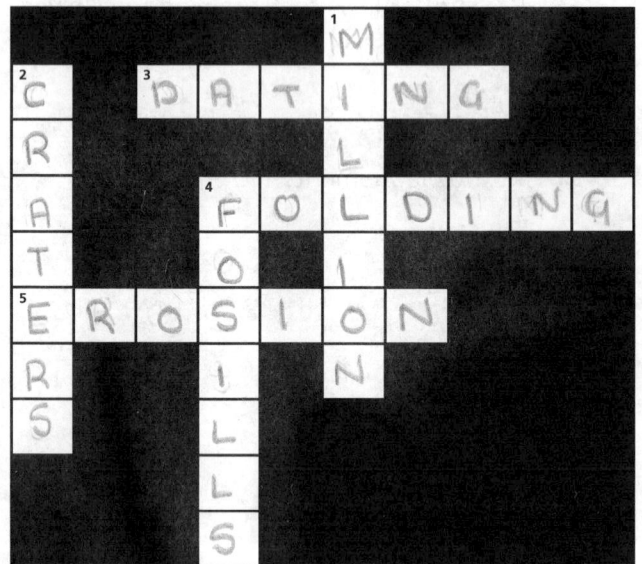

2 Label the different layers on this diagram of the Earth's structure, giving the name of, and one fact about, each layer.

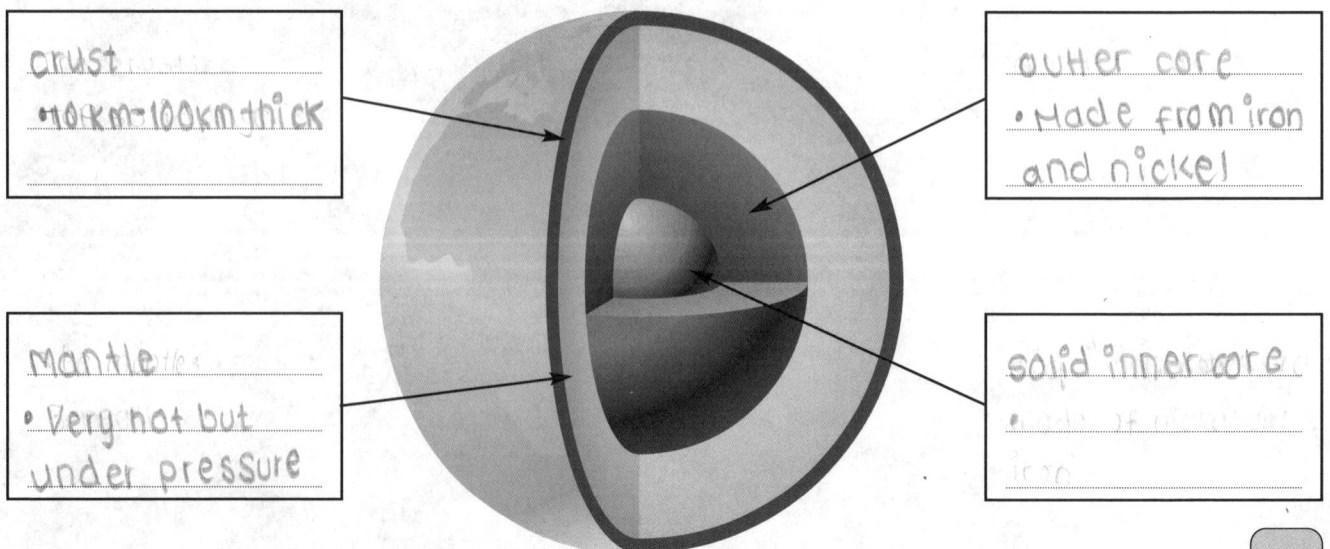

crust
•10km-100km thick

outer core
• Made from iron and nickel

mantle
• Very hot but under pressure

solid inner core

The Earth in the Universe

1 a) Who was Alfred Wegener?

A meterologist

b) Outline the theory of continental drift.

Wegener believed that the continents were once joined together but slowly drifted apart

c) What is meant by 'the jigsaw fit'?

The continents fitted together like 'jigsaws'

d) Give two reasons why Wegener's theory was not accepted by other scientists.

i) *He was not a geologist so was considered an outsider*

ii) *There was not enough evidence*

2 The Earth's crust is cracked into several tectonic plates, which float on top of the mantle.

a) Why do these plates float on top of the mantle?

They are less dense

b) What is the name given to the points where plates meet?

plate boundries

3 a) What is a geohazard?

A geohazard is a natural disastar caused by the earth e.g. earthquakes, landslides

b) Different geohazards require different measures to be taken to reduce the risks to people and property. List two different geohazards, and some of the safety measures that can be taken to reduce the risks.

i) *evacuate the area*

ii) *put up emergency services on standby*

The Earth in the Universe

1 Use the words provided to fill the gaps and complete this description of how the seafloor spreads.

oceanic   tectonic   liquid   constructive

magma   mantle   convection   continents

Just below the crust the ____mantle____ is fairly solid. However, further down it is

____liquid____ and able to move. ____convection____ currents in the mantle carry the

____tectonic____ plates, moving entire ____continents____. Where these currents cause plates

to move apart, ____magma____ rises to the surface, and new areas of ____oceanic____

crust are formed. A boundary where new crust is formed is called a ____constructive____ plate boundary.

HT

2 Explain how the magnetic strips of rock on the seafloor provide evidence for the seafloor spreading, and led to Wegener's theory of continental drift finally being accepted.

_____

_____

_____

_____

_____

3 Use the clues about tectonic theory to solve the crossword puzzle below.

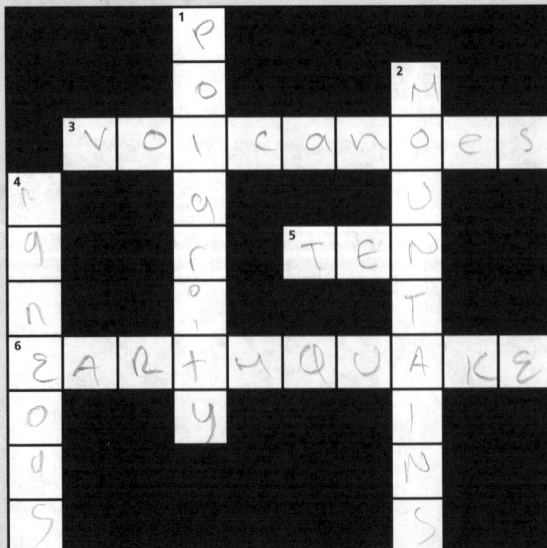

Across

3. These often form along destructive boundaries (9)
5. Plates move by about _____ cm per year (3)
6. A geohazard that occurs when built-up pressure is suddenly released (10)

Down

1. The term used to describe the direction of a magnetic field (8)
2. When rock is forced upwards in a collision, at a destructive plate boundary, these are formed (9)
4. The type of rock that forms when magma reaches the surface (7)

1 a) In addition to energy, what else is produced during nuclear fusion?

b) Explain how this proves the statement: *Our Solar System was once part of a star*.

2 a) Draw lines between the boxes to connect each of the objects to its correct definition.

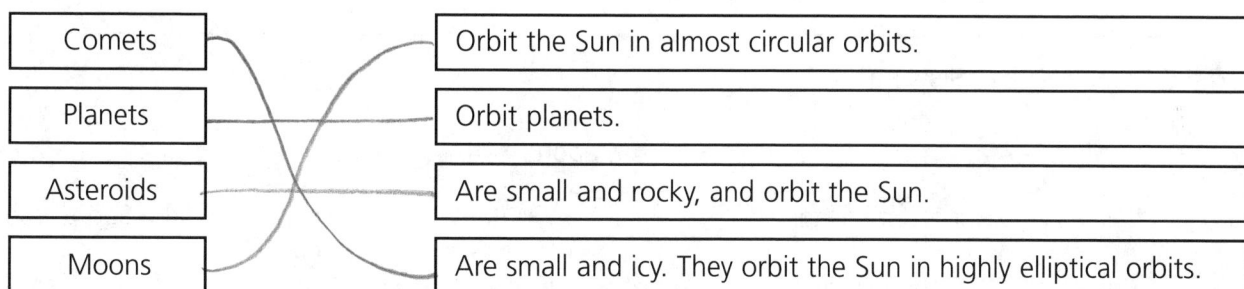

Comets	Orbit the Sun in almost circular orbits.
Planets	Orbit planets.
Asteroids	Are small and rocky, and orbit the Sun.
Moons	Are small and icy. They orbit the Sun in highly elliptical orbits.

b) Below is an image of the Sun. Draw lines around the Sun to show the shape and direction of the following orbits:

i) The Earth

ii) A comet

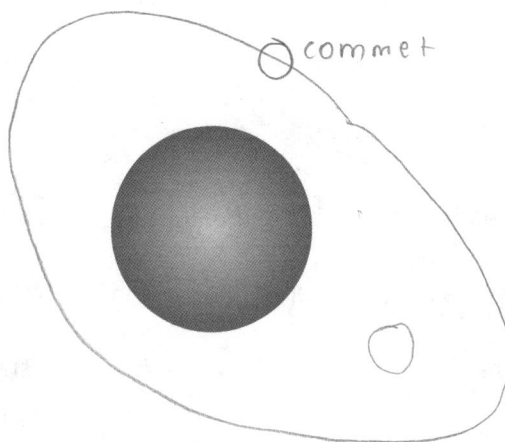

commet

HT

3 In a star, nuclear fusion joins nuclei together, releasing energy. Which nuclei are fused together?

Hydrogen and helium

4 How long ago was the Solar System formed?

5 000 million years ago

The Earth in the Universe

1 a) How old is our Sun? _500 million years old_

b) How old do scientists think the Universe is? _1400 million years old_

c) Put a (circle) around the best estimate for the number of stars in the Milky Way galaxy.

2 million   (**200 million**)   **2 billion**   **200 billion**   **2 trillion**

d) Write the following in order of size, starting with the smallest.

Star   **Planet**   **Moon**   **Galaxy**   **Universe**

Moon, planet, star, galaxy, universe

HT **2 a)** Complete the following:

Light travels at _300 000_ km/s, which is about _1 million_ times faster than sound. Even at this incredible speed it still takes _8_ minutes for light to travel from the Sun to the Earth. The light from distant galaxies will have taken millions of years to reach us.

b) *Astronomers can look back in time.* Explain, in as much detail as you can, what this statement means.

_____

_____

_____

3 Which of the following is the correct definition of a light year? Tick the correct answer.

a) The time light takes to travel a certain distance. ☐   **d)** The distance light travels each second. ☐

b) The time light takes to travel in one year. ☐   **e)** The distance light travels in a day. ☐

c) The distance light travels in one year. ☑

4 Give two methods that astronomers use to work out how far away stars are, and describe a problem for each.

a) _Relative brightness_

b) _parallax_

1 Choose the correct labels from the following list and add them to the diagram (in the boxes provided) to show the life cycle of a star. The first one has been done for you.

White dwarf   **Red giant**   **Heavy stars**   **Black dwarf**

Neutron star   **Lighter stars**   **Supernova**   **Planetary nebula**

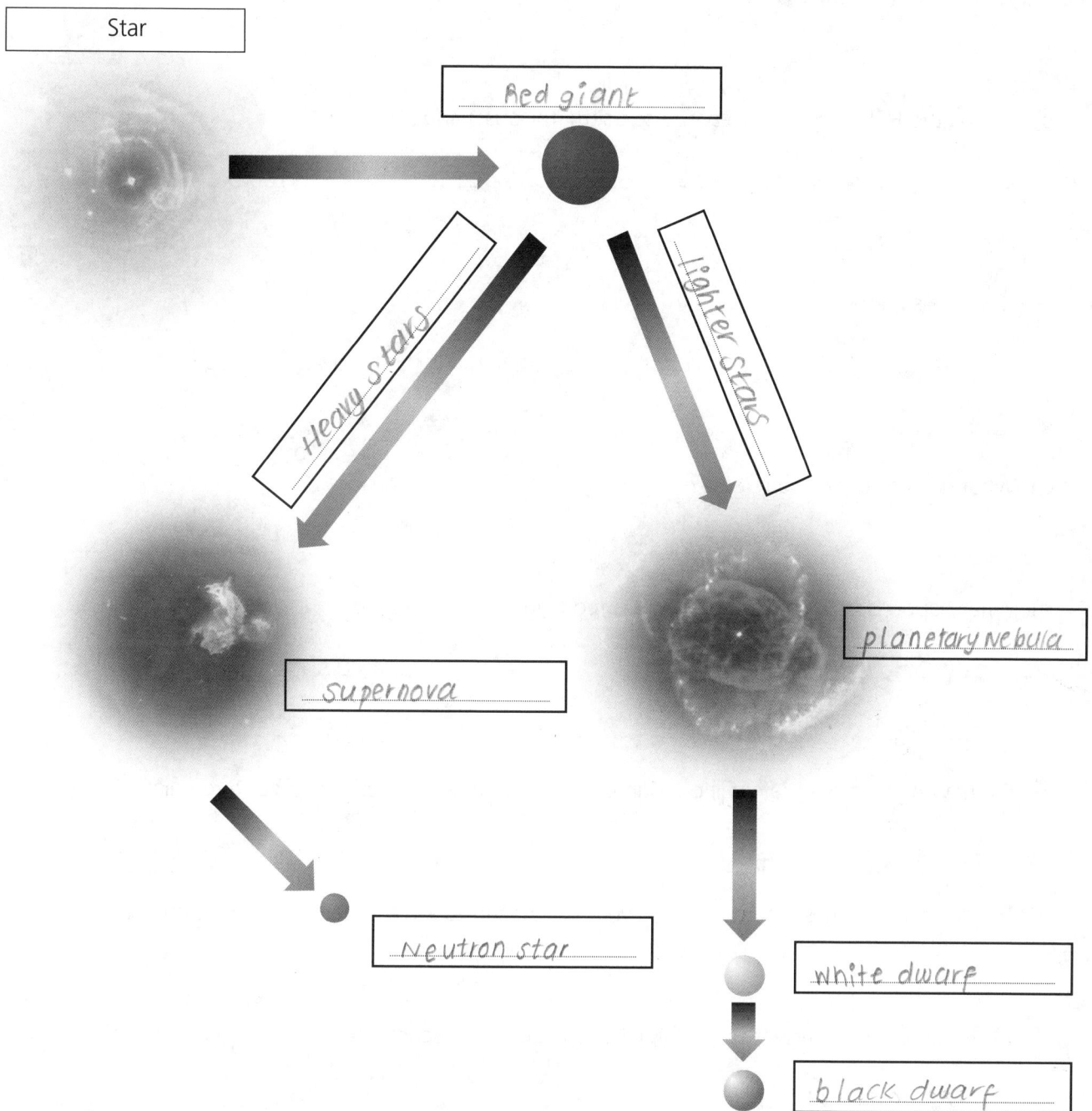

Star

Red giant

Heavy stars

Lighter stars

supernova

planetary nebula

Neutron star

white dwarf

black dwarf

The Earth in the Universe

2 Name three types of electromagnetic radiation that is given out by stars.

a) _Visible light_

b) _Ultraviolet_

c) _infrared_

3 a) What is light pollution?

Electric lights on Earth illuminate the night sky so it is difficult to see the stars

b) Why is the Hubble Space Telescope so useful for observing stars?

It is not affected by light pollution as it orbits the Earth at a height of 600km

c) How is a supernova formed?

A red giant shrinks and explodes releasing massive amounts of energy, dust and gas

d) What does a supernova release?

energy, dust, gas

e) What happens to a star as it becomes a red giant?

It will swell up becoming colder and colder

f) For how many more years will our Sun continue to shine before it becomes a red giant?

5000 million years

g) i) What determines whether a star eventually becomes a black dwarf or a neutron star?

size

ii) Based on your answer to part i), what will our Sun become at the end of its life cycle?

red giant

The Earth in the Universe

1 Complete the following sentence by circling the correct option:

If a source of light is moving away from us, the wavelengths of the light are **longer**/**shorter** than they would be if the source was stationary.

HT **2** Describe, in as much detail as you can, Hubble's Law.

3 What does the 'Big Bang' theory state? *The universe started with a huge explosion*

4 a) What does the future of the Universe depend on? *mass in the universe*

b) Give two possible outcomes for the future of the Universe and describe the conditions that would cause each one.

Outcome	Conditions that would cause it
i) *Universe will keep expanding*	*Not enough mass*
ii) *Universe will collapse*	*too much mass*

5 a) What did NASA announce that they had found in a meteorite from Mars?

Fossil of ancient life

b) What happened when other scientists studied the rock?

Offered different explanations

6 a) Why do most scientists agree that there is probably life somewhere else in the Universe?

b) Why will scientists not say for definite that there is other life out there?

Not enough evidence

The Earth in the Universe

1 a) Approximately how long ago did the dinosaurs become extinct?

65 million years ago

b) Explain, as fully as you can, why many scientists believe that the dinosaurs' extinction was caused by an asteroid collision.

_____

_____

_____

c) Describe three pieces of evidence which prove that asteroids have collided with Earth in the past.

i) _____

ii) _____

iii) _____

d) The extinction of the dinosaurs took place over a few million years. Why does this fact lead to some scientists arguing against the asteroid theory?

_____

_____

e) Why might humans have a better chance than the dinosaurs of surviving an asteroid collision?

_____

_____

_____

2 What is the difference between data and theories?

_____

_____

_____

_____

1 Use the answers to the clues below to complete this crossword.

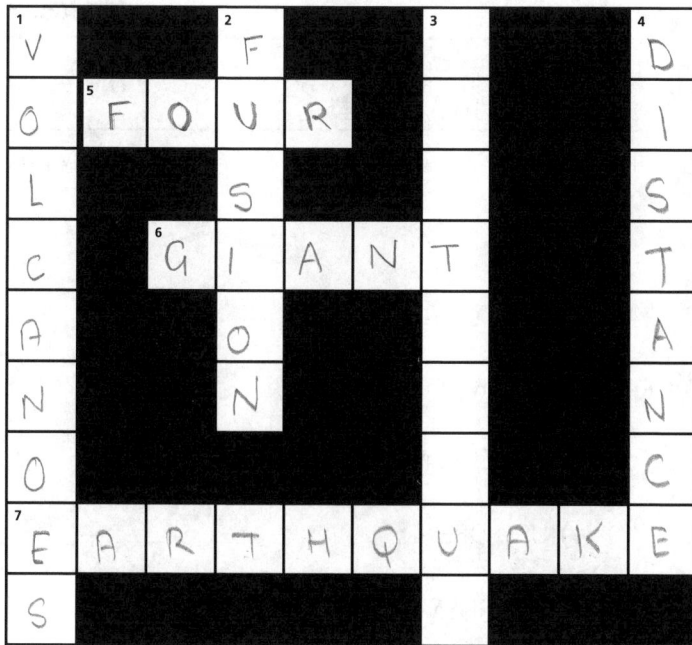

```
 1          2            3          4
 V          F                       D
 O  5 F O U R                       I
    F
 L          S                       S
 C  6 G I A N T                     T
 A          O                       A
 N          N                       N
 O                                  C
 7 E A R T H Q U A K E
 S
```

Across

5. The number of layers the Earth has (4)
6. A red _giant_ forms as a star cools (5)
7. One of these can occur when built-up pressure in the Earth's crust is suddenly released (10)

Down

1. These often form along destructive boundaries (9)
2. Nuclear _fusion_ releases energy in a star (6)
3. The first stage of a star's life (4,5)
4. A light year is the _distance_ light travels in a year (8)

2 Fill in the gaps in the following sentences, using some of the words below.

elements	thick	thin	atoms
hot	energy	cold	air

The Earth's crust is a _thin_ layer of solid rock. The mantle is a _thick_ layer of rock.

The inside of the Earth is kept _hot_ by the _energy_ released when the

atoms of radioactive _elements_ decay.

Radiation and Life

1 a) Complete the diagram below by putting the different types of radiation in the correct places.

Low frequency   High frequency

i) Radio waves	Microwaves	**ii)** infrared	**iii)** visible	Ultraviolet Rays	**iv)** x-ray	**v)** Gamma

Low energy photons   High energy photons

b) What is a photon?

packets of energy

2 a) Rearrange the following anagrams to reveal examples of radiation emitters.

i) ELCERTIC REFI   Electric fire   **ii)** OIBLEM PNEOH STMA   Mobile phone mast

iii) MOERTE LCOTNOR   Remote control   **iv)** USN DBE   sun bed

v) NUS   sun   **vi)** VT TRNSMATETRI   Tv transmitter

vii) AYRX HMAINEC   x-ray machine

b) Choose four of the emitters listed in part a) and complete the table below to explain how the waves travel.

Emitter	How the Waves Travel
i)	
ii)	
iii)	
iv)	

c) The amount of radiation received by the detector is always less than the amount transmitted by the emitter. Why is this?

Radiation and Life

1 a) Complete the following sentences.

i) The intensity of electromagnetic radiation is the _energy_ arriving at a surface _per_ _second_ .

ii) The intensity of a beam of radiation _decreases_ with distance, so the further away from the source you are, the _lower_ the intensity.

b) What two things does the intensity of electromagnetic radiation depend on?

i) _Number of photons delivered per second_

ii) _Amount of energy each packet contains_

2 When radiation is absorbed it produces heat. What does the amount of heat produced depend on?

The intensity

3 Describe, in as much detail as you can, why the intensity of a beam of radiation decreases with distance travelled.

- photons spread out as they travel so energy is more spread out

- some photons are absorbed by particles

- some photons are reflected

4 a) Describe what is meant by the term 'ionising radiation'.

- High photon energy

- break molecules into ions

b) What can ionising radiation do to molecules?

break molecules into ions

c) Name three types of ionising radiation.

i) _Ultraviolet_   ii) _X rays_   iii) _gamma rays_

Radiation and Life

1 Solve the clues about the effect of radiation on cells to complete the crossword.

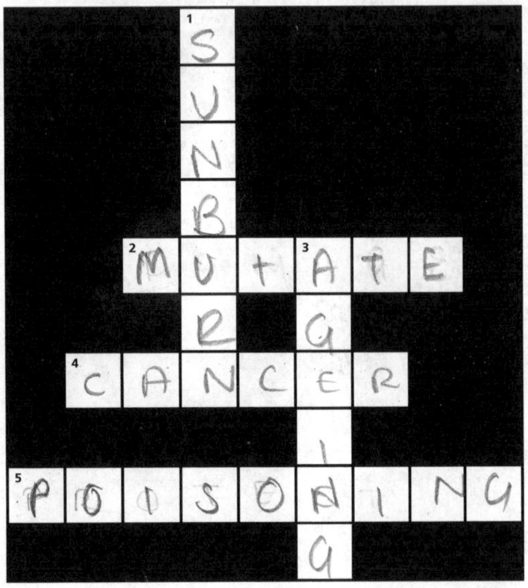
SUNBURN
MUTATE
AGEING
CANCER
POISONING

Across

2. Ionising radiation absorbed by the cells can cause them to do this (6)
4. A disease that can be caused when ionising radiation damages the cells (6)
5. Radiation .... could occur when radiation kills cells (8)

Down

1. Can be caused by the heating effect of radiation from the Sun (7)
3. Cell damage from ultraviolet light can cause .... of the skin (6)

2 How do microwaves heat materials?

Microwaves heat materials by heating water molecules which cause them to vibrate fast

3 **a)** Why is radiation protection important?

Radiation protection is important as it can damage cells in the body

b) Draw a line to connect each of the following types of radiation to its correct protection method.

X-rays	Suntan lotion
Ultraviolet light	Wire screen / thin metal
Gamma rays	Lead screens / aprons
Microwaves	Thick lead and concrete

Radiation and Life

1 Give two reasons why we would not be able to survive without energy from the Sun.

a) The place would be too cold

b) For photosynthesis

2 How does photosynthesis affect the atmosphere?

Removes carbon dioxide and releases oxygen

3 What is the ozone layer and how does it protect living organisms?

The ozone layer is a layer of gas surrounding earth that controls the amount of UV light coming in

4 a) Explain what effect greenhouse gases have on the heat escaping into space. Label the diagram alongside to illustrate your answer.

b) Name one greenhouse gas.

Methane

5 Fill in the gaps in the following sentences, using some of the words listed below.

| warmer | greenhouse effect | emits | absorb | global warming |
| gases | cooler | liquids | radiation | |

The Earth _emits_ electromagnetic _radiation_ into space, but there are _gases_ in the atmosphere which _absorb_ some of this _greenhouse effect_ This keeps the Earth _warmer_ than it would otherwise be, and is known as the _global warming_ .

Radiation and Life

1 a) Add the labels below to the correct arrows on the diagram.

Respiration	Respiration	Respiration
Photosynthesis	Death of plants	Eating of plants

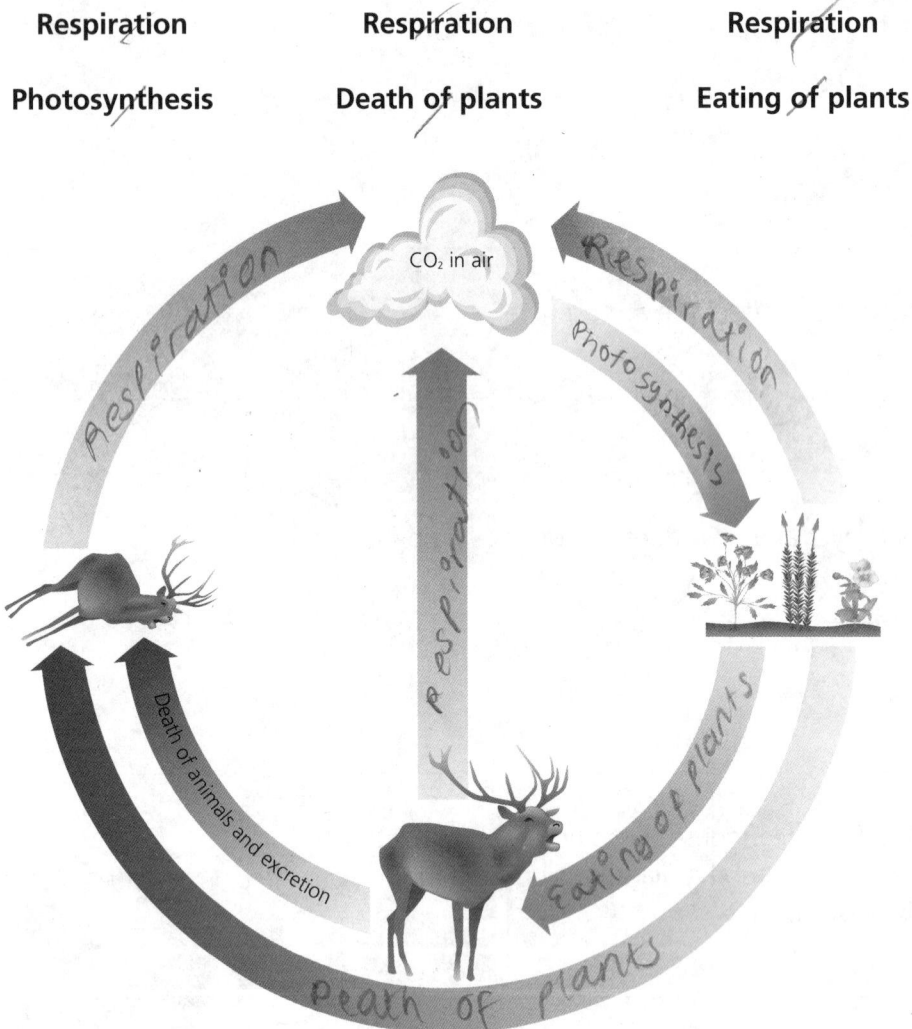

CO₂ in air

Handwritten labels on diagram: Respiration, Respiration, Photosynthesis, Death of animals and excretion, Respiration, Eating of plants, Death of plants

b) i) Which process removes carbon dioxide from the atmosphere?

~~Respiration~~ green plants

ii) What types of organism take part in this process?

plants and animals

iii) Explain how the carbon from carbon dioxide in the atmosphere ends up in organisms further up the food chain and is then returned to the atmosphere.

2 The amount of carbon dioxide in the atmosphere has remained constant for thousands of years. Explain how.

The levels of CO_2 are continuously recycled.

3 Explain how burning fossils fuels and deforestation affect the overall level of atmospheric carbon dioxide.

4 a) What is global warming?

b) List three possible effects of global warming.

i)

ii)

iii)

5 Why do climatologists believe that human activity is responsible for global warming?

Radiation and Life

1 Gamma rays can be used to kill cancer cells but they can also cause harm to people who are exposed to them.

a) Use this statement to explain what we mean when we talk about weighing up the risk against the benefit.

....

....

....

....

b) Think of, and explain, another use of radiation where the risk has to be carefully assessed.

....

....

....

....

2 A newspaper said that in a recent study of ten mobile phone owners, one person was found to have a brain tumour. The newspaper said this proves that microwaves from mobile phones have a 10% chance of causing a brain tumour. Scientists say it was a very poor study and does not give enough information.

Explain why the scientists think this is a very poor study. (Hint: think about sample size and what other information the scientists might be referring to).

....

....

....

....

....

....

3 a) Many activities have a risk associated with them. What do people mean when they talk about 'weighing up the risks'?

....

....

b) Suggest two benefits of mobile phones.

i) ....

ii) ....

Radiation and Life

1 Too much sunlight can cause skin cancer, but sunlight can also be good for you.

a) Which of the following are possible health benefits of sunlight? Put a tick in the correct box(es).

i) It is a source of vitamin B. ☐ **ii)** It is a source of vitamin D. ☐

iii) A tan makes people look good. ☐ **iv)** It can help to prevent SAD (Seasonal Affective Disorder). ☐

b) Which of the following are personal, rather than medical, reasons for sunbathing? Put a tick in the correct box(es).

i) It is a source of vitamin B. ☐ **ii)** It is a source of vitamin D. ☐

iii) A tan makes people look good. ☐ **iv)** It can help to prevent SAD (Seasonal Affective Disorder). ☐

2 a) Give an example of an activity that poses a risk but that people do anyway. State what the risk might be.

Activity: _____ Risk: _____

b) Using your example above, suggest why people do this activity in spite of the risk.

_____

_____

_____

HT

3 a) What is the difference between perceived risk and actual risk?

_____

_____

b) Suggest two factors that could influence perceived risk.

i) _____ **ii)** _____

c) i) What does 'ALARA' stand for?

_____

ii) Give an example of when the ALARA principle would be used as a guideline for risk management.

_____

_____

Radiation and Life

1️⃣ Solve the clues below to fill in the crossword.

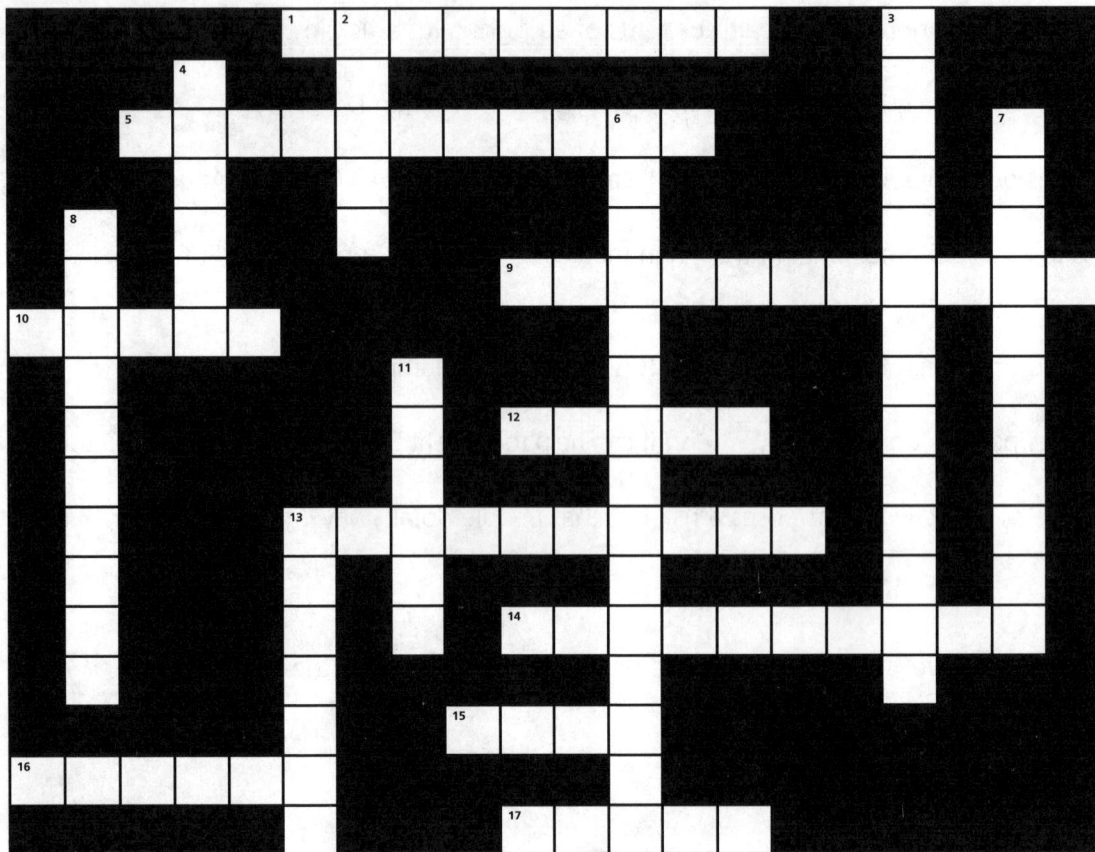

Across

1. Electromagnetic radiation given out by a supernova (5,4)
5. The type of radiation that can cause skin cancer (11)
9. These organisms break down dead plant and animal material (11)
10. These can be damaged by radiation (5)
12. The _____ layer absorbs ultraviolet radiation from the Sun (5)
13. Used for mobile phone communication (10)
14. High-energy photons can cause this (10)
15. The type of energy produced when radiation is absorbed (4)
16. A packet of energy (6)
17. The continual recycling of carbon compounds is called the carbon _____ (5)

Down

2. A principle applied when using radiation (5)
3. The process by which plants remove carbon dioxide from the atmosphere (14)
4. _____ warming is happening because of an increase in greenhouse gases (6)
6. The _____ spectrum (15)
7. A link between two variables, not necessarily cause and effect (11)
8. The _____ effect helps to keep the Earth warm (10)
11. This can occur when radiation causes DNA to mutate (6)
13. A greenhouse gas (7)

Radioactive Materials

1 Fill in the blanks in the following passage:

There are many different types of atom; an element contains only .... type of ....

Different elements contain different .... All atoms contain a .... and ....

The .... is made up of .... and neutrons.

2 Add labels to the diagram below to identify the nucleus and the electrons.

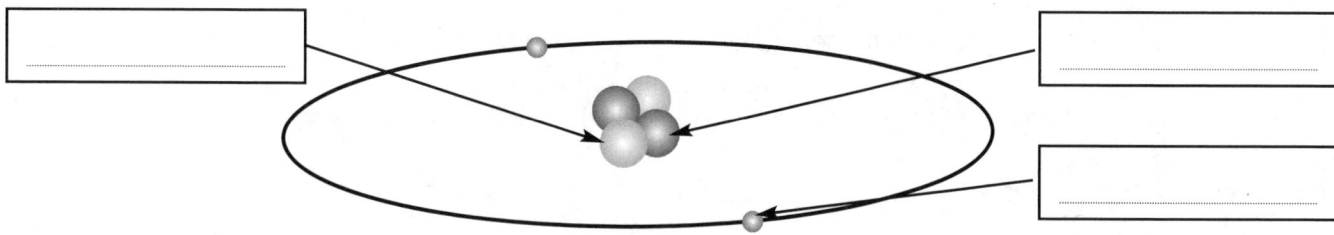

3 What is the term used to describe elements that give out ionising radiation?

....

4 Complete the table below.

Type of Radiation	Ionising Power	Penetrating Power
Alpha	a)	Stopped by a thin sheet of paper or a few cm of air.
b)	Reasonably ionising	c)
Gamma	d)	e)

5 In terms of the different elements, what is the significance of the number of protons in the nucleus?

....

....

6 What is an isotope?

....

....

Radioactive Materials

HT **1** For each of the following questions, place a tick alongside the correct answer.

a) Ionising radiation is emitted when…

i) two atoms react ☐ **ii)** an element vaporises ☐

iii) the nucleus of an unstable atom decays ☐ **iv)** an atom is ionised. ☐

b) During alpha decay…

i) an electron is ejected from the nucleus ☐ **ii)** no new elements are formed ☐

iii) a new element is formed ☐ **iv)** the atom dissolves. ☐

c) During beta decay…

i) a proton turns into a neutron and an electron ☐ **ii)** electromagnetic radiation is emitted ☐

iii) the atom loses a proton ☐ **iv)** a neutron turns into a proton and an electron. ☐

d) During gamma decay…

i) an electron is ejected from the nucleus ☐ **ii)** no new elements are formed ☐

iii) a new element is formed ☐ **iv)** surplus energy is absorbed by the nucleus. ☐

2 Why does radioactive decay occur? (You should refer to stability in your answer.)

_____

_____

3 **a)** Which of the following radiations is the odd one out? Tick the correct answer.

i) Alpha decay ☐ **ii)** Beta decay ☐ **iii)** Gamma decay ☐

b) Explain your answer to part a).

_____

_____

_____

_____

4 Briefly describe each of the following:

a) An alpha particle.

_____

b) A beta particle.

_____

c) A gamma ray.

_____

5 a) What is background radiation?

_____

b) Name three sources that contribute to background radiation.

i) _____ **ii)** _____ **iii)** _____

c) Should we be concerned about exposure to background radiation? Explain your answer.

_____

_____

6 What is the activity of a radioactive substance a measure of?

_____

_____

7 What is the definition of 'half-life'? Tick the correct answer.

a) The time it takes for the activity of a radioactive material to halve. ☐

b) The time it takes for the weight of a radioactive material to halve. ☐

c) The time it takes for a radioactive atom to decay by 50%. ☐

d) The time it takes for the strength of a radioactive substance to halve. ☐

8 *A half-life is a very short time.* Is this statement accurate? Explain your answer.

_____

Radioactive Materials

9 All radioactive substances become less radioactive as time passes. At what point would a radioactive substance be considered safe?

....

....

HT **10** A radioactive material has the following activity:

Time (years)	0	10	20	30	40	50	60	70
Activity (Bq)	1000	650	400	248	150	85	60	52

a) Plot the results on the graph below.

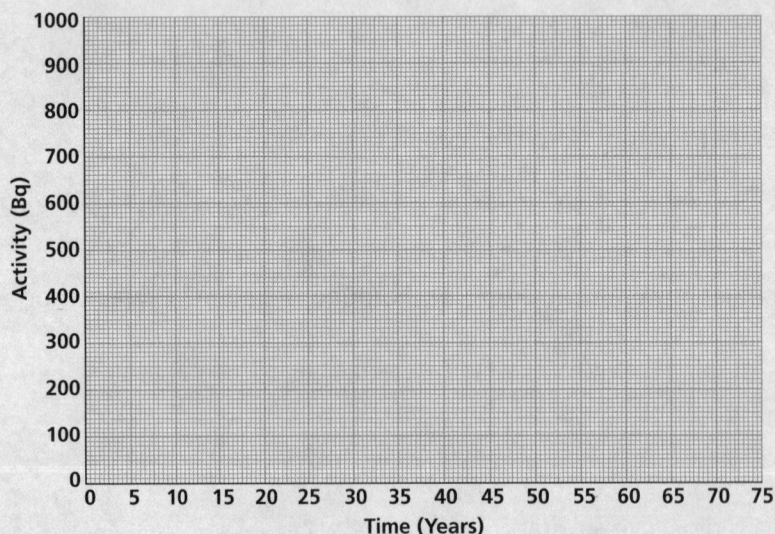

b) Use the graph to find the half-life of the material.

....

c) If a sample of the same material started with an activity of 4000 counts per minute, how many years would it take for the activity to fall to 1000 counts per minute?

....

11 A radioactive substance has a count rate of 1600 Bq and a half-life of 2000 years. What would the count rate be after…

a) 2000 years? ....

b) 4000 years? ....

c) 6000 years? ....

Radioactive Materials

1 a) Alpha radiation is most dangerous if the source is inside the body. Why is this so?

....

....

b) Why is alpha radiation less dangerous if the source is outside the body?

....

....

2 Gamma radiation is the most penetrating, but is the least damaging to the body. Why is this so?

....

....

3 Why is beta radiation the most dangerous if the source is outside the body?

....

....

4 What is a sievert?

....

5 Complete the table below which summarises how different types of radiation are used to treat cancer:

Radiation Type	How it is Used	Dangers and Safety Measures
High energy gamma ray	a)	b) ....
Beta rays from radioactive iodine	c)	The beta radiation could damage healthy cells as well as the cancer cells. Therefore, the ALARA principle is applied.

6 Describe two other practical uses of radiation.

....

....

....

....

Radioactive Materials

1 Why is electricity described as a secondary energy source?

....

....

2 Why is using electricity less efficient than using a primary energy source?

....

....

....

3 On the flow diagram below, complete the labels showing the main stages of electricity generation from fossil fuels.

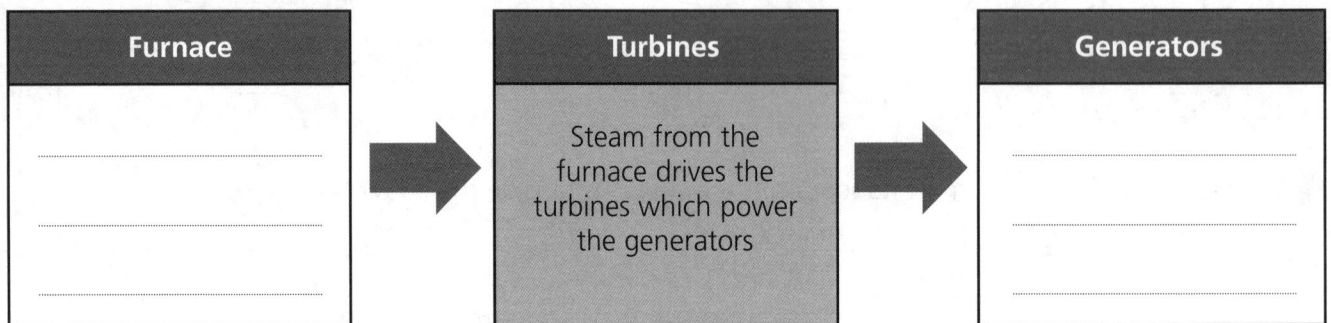

Furnace	Turbines	Generators
	Steam from the furnace drives the turbines which power the generators	

4 What environmental issues are associated with burning fossil fuels?

....

....

....

5 Where does the energy come from in a nuclear power station?

....

....

6 a) Using nuclear power to generate electricity produces nuclear waste. Use a line to join each type of waste to its description.

High-level waste	Fairly radioactive, and remains radioactive for thousands of years.
Intermediate-level waste	Only slightly radioactive; stored in landfills.
Low-level waste	Very radioactive, but only small amounts are produced and it does not remain radioactive for very long.

b) Which of the three types of waste do you think is the worst from an environmental point of view? Explain your answer.

....

....

Radioactive Materials

1 Energy is lost when electricity is generated. What happens to this energy?

....

....

2 Suggest two stages of the generation process where energy is lost.

a) .... **b)** ....

3 Choose two renewable energy resources and summarise the key points about each method in the table below:

Energy Resource	How it Works	Power Output	Effect on Environment

4 The inhabitants of a small island off the coast of Scotland have decided that they need their own power supply. The island has a population of 5000, is windy most of the year and is surrounded by water. A large amount of the land is unpopulated. The inhabitants are considering a few different options. Use the information provided in the table below to suggest a suitable generation method for them to use. Explain your answer.

Wind Power	Nuclear Power	Wave Power	Coal Power
• Low set-up cost. • Low power output. • Requires wind. • Some residents may object to appearance of wind turbines.	• Very high set-up cost. • High power output. • Very reliable. • Produces nuclear waste.	• Low/medium set-up cost. • Medium power output. • Needs large areas of sea. • Still experimental and prone to break down.	• High set-up cost. • High power output. • Needs large amount of coal daily. • Very reliable.

....

....

....

....

Radioactive Materials

HT

1 Complete the crossword below about nuclear reactors.

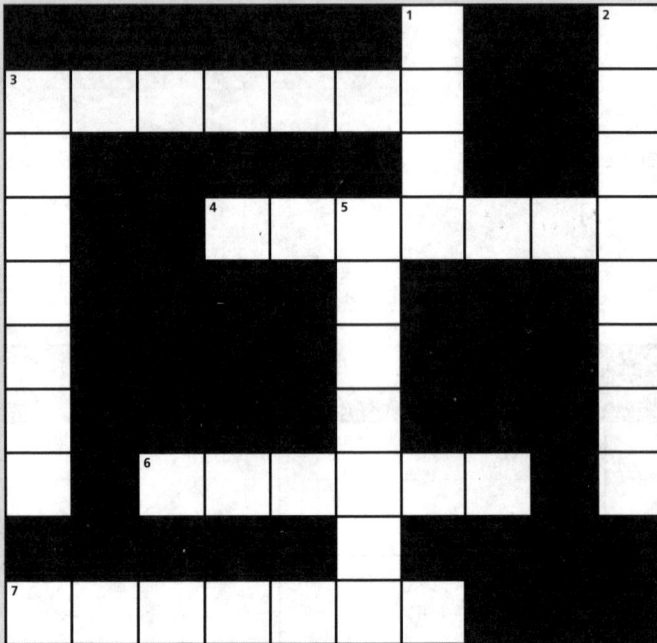

Across

3. The name given to material that will sustain a chain reaction (7)
4. Rods used to prevent the reaction getting out of control (7)
6. Large amounts of this are released in a fission reaction (6)
7. The liquid pumped through a nuclear reactor and used in the heat exchanger (7)

Down

1. The type of energy produced in the reactor core during fission reactions (4)
2. The bars in a reactor where the reaction takes place (4,4)
3. This takes place in the nucleus of the atom and forms new elements (7)
5. One of these is absorbed by an unstable nucleus to trigger a fission reaction (7)

2 **a)** Explain what a chain reaction is.

....

....

b) Why is it important to control a chain reaction?

....

....

c) In a nuclear reactor, how is the chain reaction controlled?

....

....

d) i) Which radioactive element is used to start the chain reaction in a nuclear reactor?

....

ii) Which two new elements are formed when fission takes place?

....

Radioactive Materials

1. For each definition below, find the relevant word(s) in the wordsearch. The answers may run horizontally, vertically, diagonally, forwards or backwards.

D	N	L	C	H	E	G	I	C	N	L	P	Y	B	X	O	L	T
A	O	J	N	T	H	M	C	A	O	A	C	T	A	R	S	E	J
T	I	P	R	O	T	O	N	S	N	W	Q	V	C	E	T	W	H
A	T	G	A	M	M	A	V	A	R	I	A	D	K	T	N	R	B
H	C	O	Q	Y	M	F	R	T	E	A	U	M	G	H	E	E	G
V	A	R	I	T	B	L	E	S	N	T	F	O	R	T	M	P	D
Q	E	N	C	W	G	X	C	S	E	G	A	N	O	N	E	U	S
U	R	E	E	A	S	A	M	E	W	V	G	T	U	M	L	E	C
R	L	E	M	U	V	O	E	R	A	W	C	R	N	P	E	V	T
E	A	L	P	Z	T	Y	A	T	B	V	S	O	D	E	V	I	R
Y	K	E	O	A	I	R	T	I	L	T	H	A	R	M	V	T	U
C	I	C	U	T	H	N	O	O	E	C	A	X	C	V	P	C	E
T	M	T	N	H	E	K	P	N	D	P	C	Z	B	N	Y	A	V
F	E	R	D	D	F	J	R	U	S	N	I	Q	W	L	O	O	A
N	H	I	I	V	E	E	B	D	F	R	M	C	X	Z	A	I	L
T	C	C	R	M	F	Y	I	P	G	S	E	T	Y	E	C	D	U
S	C	I	T	F	R	Y	V	S	I	E	V	E	R	T	L	A	E
A	T	T	I	Y	P	B	M	O	L	E	C	U	L	E	S	R	Y
J	I	Y	P	R	O	P	S	R	T	I	E	S	T	W	Y	I	O

a) A nucleus is made from _____ and neutrons
b) An isotope has a different number of these
c) A type of ionising radiation that does not change the element
d) A measure of a radiation dose, depending on the amount and type of radiation received
e) A secondary energy source
f) Coal is an example of this type of energy source
g) Any activity contains a certain risk of _____ or _____
h) Uranium, which releases energy as it decays, is a _____ element
i) A general name for radiation that is all around us

© Lonsdale PHYSICS WORKBOOK – Revision Guide Reference: Page 29 31

Explaining Motion

1 What is the difference between speed and velocity?

_____

_____

2 If a car has a velocity of +10m/s, what is its velocity when it is travelling in the opposite direction at the same speed?

_____

3 A man walks +10 metres and then -5 metres. What is...

a) the total distance travelled?

_____

b) the man's distance from the start point?

_____

4 Write down the formula for calculating speed.

_____

5 Complete the following table:

Speed (m/s)	Distance (m)	Time (s)
15	45	**a)**
6	60	**b)**
c)	100	10
d)	300	60
25	**e)**	4
30	**f)**	20

6 What is the difference between average speed and instantaneous speed?

_____

7 A child playing in the playground runs at 4m/s for 3 seconds, stops for 5 seconds and then runs at 4m/s for 2 seconds. Calculate…

a) the distance travelled in the first 3 seconds ....

b) the distance travelled in the next 5 seconds ....

c) the distance travelled in the last 2 seconds ....

d) the total distance travelled ....

e) the total time taken ....

f) the average speed. ....

8 A car travels 15 miles from Sheffield to Doncaster in a time of 30 minutes. The car stops in Doncaster for 1 hour, and then returns to Sheffield, again taking 30 minutes to travel the 15 miles.

a) How fast is the car travelling from Sheffield to Doncaster?

....

b) What is the average speed for the entire 2 hours?

....

9 Sketch the following distance–time graphs.

a) A woman remaining stationary 5m from a starting point (0).

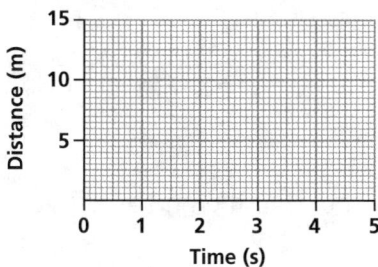

b) A runner travelling at a constant speed of 5m/s.

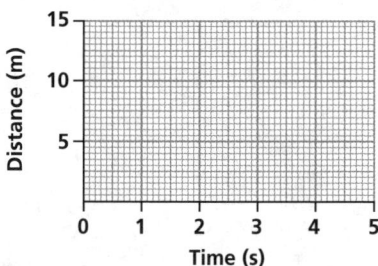

Explaining Motion

1 The graph below shows a car's journey.

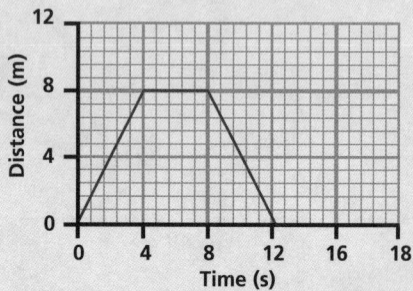

a) What is the average velocity of this journey? ....

b) What is the average speed of this journey? ....

2 On a distance–time graph what does a curved line indicate?

....

3 Sketch a distance–time graph showing a jogger starting from rest and running 7m in 5 seconds at an increasing speed.

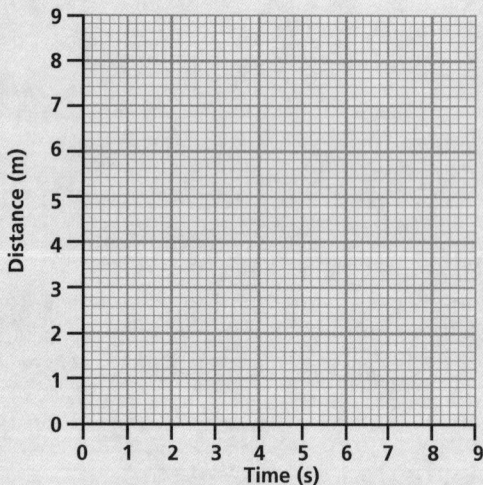

4 Sketch a distance–time graph of a cyclist travelling at 4m/s for 2 seconds and then gradually slowing down to come to rest 6m from the start position in 4 seconds.

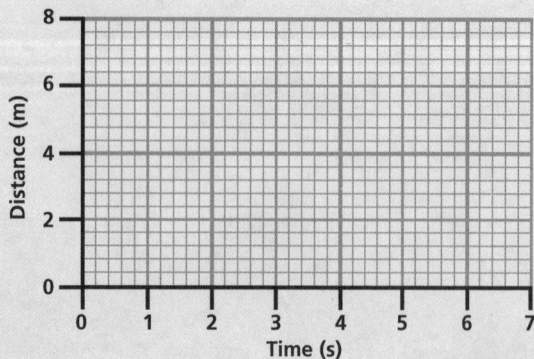

Explaining Motion

1 What does the slope of a velocity–time graph represent?

_____

2 On the graphs below, sketch the following velocity–time graphs, illustrating…

a) an object at rest

b) an object travelling at a constant velocity of 5m/s

c) an object accelerating at a constant rate of 5m/s²

d) an object decelerating at a constant rate from 5m/s² to rest.

a)

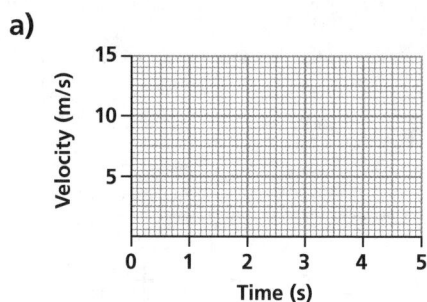

b)

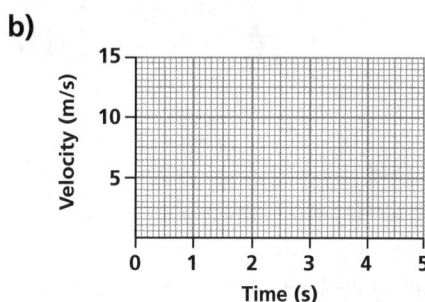

c)

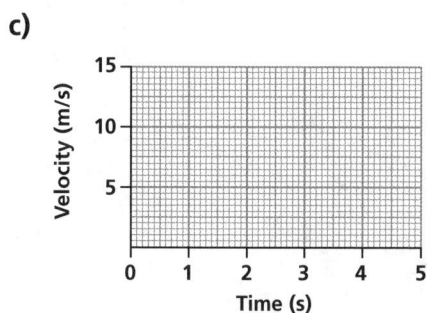

d)

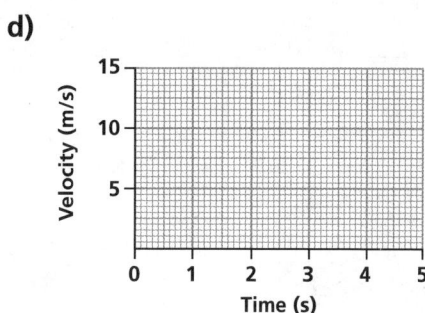

3 Velocity–time graphs are used as tachographs to record the journey of a lorry. Describe the journey depicted by the tachograph below.

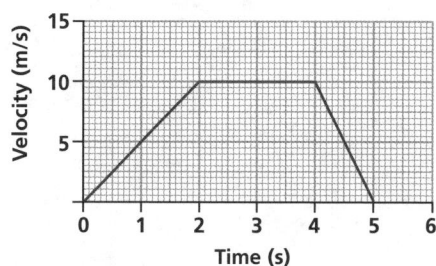

_____
_____
_____
_____

Explaining Motion

1 How does a force occur?

....

2 Complete the table below to give the names and descriptions of four forces.

Name of Force	Description of Force
a)	Acts to slow things down when two surfaces rub against each other.
Air Resistance	**b)** ....
c)	Pushes up on the bottom of a cup sitting on a table and stops the cup sinking into the table.
Gravity	**d)** ....

3 Whenever something exerts a force it experiences an equal and opposite force. Explain how this principle is used in jet engines.

....

....

....

....

4 What is the resultant force?

....

5 The following diagrams illustrate a car travelling at 30mph. Calculate the resultant force and state what happens to the motion of the car.

a) 1000N ←  → 4000N

....

b) 2000N ←  → 0N

....

c) 1000N ←  → 1000N

....

6 How do forces such as friction and reaction occur?

....

....

7 Explain the forces at work when an object is resting on a surface.

....

....

....

8 A car parked on a hill experiences a force of 5000N from gravity trying to pull it down the hill. Which of the following is the value of friction?

a) Zero ☐   **b)** Less than 5000N ☐

c) 5000N ☐   **d)** More than 5000N ☐

9 The friction on a car travelling at a constant speed is 400N and the air resistance is 300N. Which of the following is the correct value for the force from the engine?

a) Zero ☐   **b)** 300N ☐   **c)** 400N ☐

d) 700N ☐   **e)** More than 700N ☐

10 A car with a weight of 10 000N is parked on a horizontal surface. What is the force of friction caused by the brakes?

a) Zero ☐   **b)** Less than 10 000N ☐

c) 10 000N ☐   **d)** More than 10 000N ☐

11 A car with a weight of 10 000N is parked on a horizontal surface. What is the value of the reaction force?

a) Zero ☐   **b)** Less than 10 000N ☐

c) 10 000N ☐   **d)** More than 10 000N ☐

12 A car with a weight of 10 000N is travelling at a constant speed on a horizontal surface. What is the value of the reaction force?

a) Zero ☐   **b)** Less than 10 000N ☐

c) 10 000N ☐   **d)** More than 10 000N ☐

Explaining Motion

1 Which two physical properties does momentum depend on?

a) _____   b) _____

2 Complete the following table.

Momentum (kg m/s)	Mass (kg)	Velocity (m/s)
a)	50	10
b)	1000	30
3000	c)	30
800	d)	2
100	e)	50
20 000	f)	20

3 When a force causes a change in momentum what two things does the size of the change in momentum depend on?

a) _____   b) _____

4 a) A car travelling at 30m/s has a mass of 1000kg. What is its momentum?

_____

b) The car applies its brakes for 5 seconds. The brakes exert a constant force of 2000N. Assuming no other forces are involved, what is the change of momentum?

_____

c) What is the new momentum of the car?

_____

d) What is the new velocity of the car?

_____

5 A motorcycle with a mass of 500kg is travelling at 20m/s. If the rider applies the brakes for 2 seconds, and the brakes exert a constant force of 5000N, what is the new momentum of the motorcycle?

_____

6 A car of mass 1500kg has a momentum of 22 500kg m/s. What is its velocity?

_____

Explaining Motion

1 During a collision the occupants of a vehicle undergo a change in momentum and experience the force of the collision. Using the idea of momentum…

a) explain why the occupants of the car are more likely to suffer serious injuries when the vehicle is travelling faster

....

....

b) explain why the occupants of the car are more likely to suffer serious injuries if the collision takes place over a shorter period of time.

....

....

2 Crumple zones are a very important car-safety feature. How does a crumple zone reduce the injuries sustained by the occupants of the vehicle?

....

....

3 A car travelling at 30m/s is involved in an accident and comes to a halt in 0.1 seconds. A passenger in the car has a mass of 50kg. Calculate…

a) the momentum of the passenger before the accident

....

b) the momentum of the passenger after the accident

....

c) the change in momentum of the passenger

....

d) the force acting on the passenger during the accident.

....

....

Explaining Motion

4 Are the following statements **true** or **false**?

a) An object has more kinetic energy if it has a greater mass. _____

b) An object has more kinetic energy if it is travelling faster. _____

c) An object travelling in space has no kinetic energy. _____

d) Doubling the mass of an object doubles its kinetic energy. _____

e) Doubling the speed of an object doubles its kinetic energy. _____

5 A moving object has kinetic energy. When a moving object comes to a halt, what has happened to its kinetic energy?

_____

_____

6 A man of 70kg is running at 5m/s. Calculate his kinetic energy.

_____

7 a) What is gravitational potential energy?

_____

_____

b) How much energy is needed to lift a 100N weight to a height of 5m?

_____

_____

c) What happens to an object's gravitational potential energy when it falls?

_____

HT 8 A 1kg book falls 1m off a table. How much kinetic energy does it have when it hits the floor?

_____

9 Complete the following sentence:

When work is being done _____ is being transferred. When work is done by an object it _____ energy and when work is done on an object it _____ energy.

10 How much work is done by a crane when it raises a 1200kg car to a height of 5m?

....

....

11 The engine of a train travelling at a constant speed provides a constant driving force of 50 000N in order to overcome resistive forces.

How much work is done by the engine in travelling a distance of 20km?

....

....

....

12 A 50kg skydiver jumps out of an airplane from a height of 2000m.

a) By the time she reaches the ground how much work has been done on her by gravity?

....

....

b) If all of the work done was used to increase her kinetic energy, how much kinetic energy would she have when she reached the ground?

....

c) Reaching the ground with such a large amount of kinetic energy would be fatal. In reality she reaches the ground with much less kinetic energy. Most of the work done by gravity is used up in overcoming another force.

What is the name of this force?

....

HT **d)** With what speed would she reach the ground?

....

....

Electric Circuits

1 Complete the following sentences from the words below.

Static electricity occurs when _____ _____ builds up on an object. This can occur when

_____ materials are rubbed against each other.

Electrons carry a _____ charge, therefore, if an object gains electrons it becomes _____

charged. For an object to become _____ charged it must _____ _____ .

positively   **negative**   **lose**   **charge**   **insulating**

electrons   **negatively**   **electrical**

2 For each of the following questions, tick the correct answer.

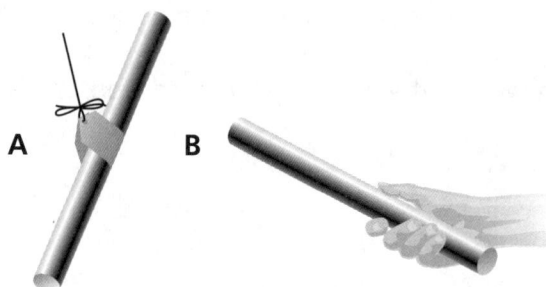

A   B

a) If both rods A and B are uncharged...

   i) nothing will happen ☐   **ii)** they will repel each other ☐   **iii)** they will attract each other ☐

b) If rod A is positively charged and rod B is negatively charged...

   i) nothing will happen ☐   **ii)** they will repel each other ☐   **iii)** they will attract each other ☐

c) If both rods A and B are positively charged...

   i) nothing will happen ☐   **ii)** they will repel each other ☐   **iii)** they will attract each other ☐

d) If both rods A and B are negatively charged...

   i) nothing will happen ☐   **ii)** they will repel each other ☐   **iii)** they will attract each other ☐

e) If rod A is uncharged and rod B is positively charged...

   i) nothing will happen ☐   **ii)** they will repel each other ☐   **iii)** they will attract each other ☐

3 A car is being spray-painted. Explain how the paint is evenly spread on the car.

_____

_____

_____

Electric Circuits

1 a) Briefly explain what an electric current is.

....

b) In what unit is current measured?

....

2 Explain why metals are good conductors of electric current.

....

....

....

3 Complete the table below.

Name	Symbol	Description
Cell	a)	Provides energy to 'push' the current around the circuit.
b)	c)	A group of cells acting together.
Filament lamp	d)	e)
f)	Ⓐ	g)
h)	Ⓥ	i)
Switch (closed)	j)	k)
l)	m)	A component whose resistance cannot be altered.

4 Explain the difference between alternating current and direct current. ....

....

5 What does the amount of current flowing through a circuit depend on? ....

....

6 What is potential difference? ....

....

7 How does the potential difference affect the amount of current that flows through a component?

....

Electric Circuits

1 What do we mean by the resistance of a component?

....

....

....

2 What effect does adding resistors have on the total resistance when they are added **a)** in series, and **b)** in parallel?

a) ....

....

b) ....

....

3 Explain how an electric current can cause a wire to melt.

....

....

....

4 Complete the table below.

Potential Difference (V)	Current (A)	Resistance (Ω)
20	2	a)
12	3	b)
32	1.5	c)
15	6	d)
8	3	e)

5 Complete the table below.

Potential Difference (V)	Current (A)	Resistance (Ω)
a)	1.5	100
b)	6	6
8	c)	20
16	d)	52

Electric Circuits

1 Sketch a graph showing the relationship between current and potential difference for a resistor at a constant temperature.

2 Complete the following sentences.

Some components have a resistance that will change depending on the environmental conditions. A

_____ dependent resistor has a resistance that _____ when the light intensity increases.

The symbol for an LDR is _____ . A thermistor has a resistance that _____ when

temperature _____ .

HT

3 What happens to **a)** the total potential difference, and **b)** the current, when batteries are added in parallel?

a) _____

b) _____

4 What is the difference between a series circuit and a parallel circuit?

_____

_____

5 Sketch a diagram of **a)** a series circuit, and **b)** a parallel circuit, which has 1 battery and 2 bulbs.

a)

b)

6 A circuit has 1 battery and 2 resistors in parallel. Resistor A has a resistance of 2 ohms and resistor B has a resistance of 1 ohm. If the battery has a 3 amp current flowing through it, what is the current flowing through resistors A and B?

Resistor A: _____   **Resistor B**: _____

Electric Circuits

7 In a parallel circuit why do components with a lower resistance have a larger current running through them?

....

....

....

8 For the following diagrams, complete the readings on the ammeters and voltmeters.

a)

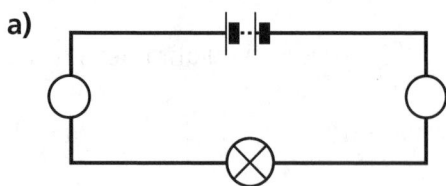

b)

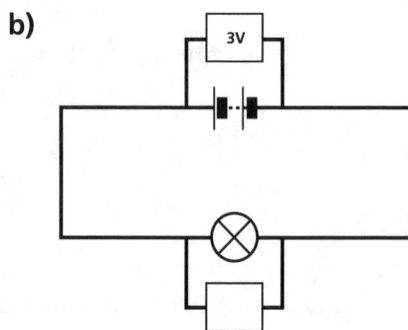

c)

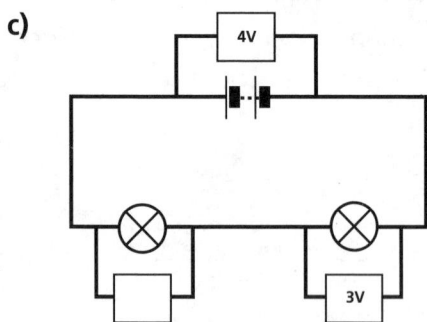

d)

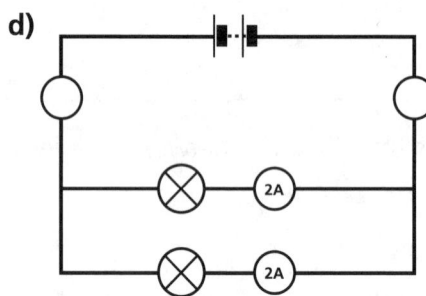

e)

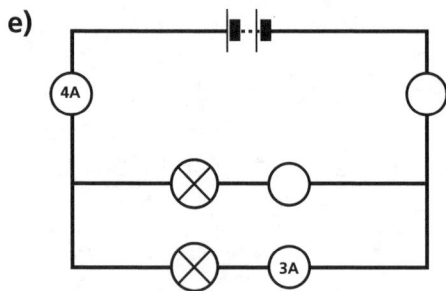

HT **9** In a series circuit with two resistors, why does the one with the larger resistance end up with a greater share of the potential difference supplied by the battery?

....

....

....

Electric Circuits

1 Moving a magnet into a coil of wire can induce an electric current in the wire. Give two ways in which the direction of the current could be reversed.

a) _____

_____

b) _____

_____

2 a) How much current would flow if the magnet is held stationary inside the coil of wire?

_____

b) Explain your answer to part a).

_____

_____

_____

3 An electric generator uses a rotating magnet inside a coil of wire. Suggest a reason why generators do not use a rotating coil and a fixed wire.

_____

_____

_____

4 Suggest three ways of increasing the size of the induced voltage.

a) _____

b) _____

c) _____

HT **5** a) Describe how the induced voltage across a coil changes during each revolution.

_____

_____

_____

b) What kind of current is produced? _____

Electric Circuits

1 Power is the rate of energy transfer. What does this mean?

....

....

2 a) What is the formula for calculating power?

....

b) When connected to a mains supply of 220 volts, an electric iron has a 10 amp current flowing through it. Calculate the power of the iron.

....

c) If the same iron is connected to a supply of 110 volts, the iron has a 5 amp current flowing through it. How much power does the iron use now?

....

HT **3** A 3kW heater is connected to a supply of 220 volts. Calculate the current.

....

4 What are transformers used for? ....

....

HT **5 a)** A step-down transformer for mobile phones converts the 240 volts mains supply to 12 volts. If the primary coil has 2000 turns, how many turns are there on the secondary coil?

....

....

b) A transformer has 40 turns on the primary coil and 200 turns on the secondary coil. If the input voltage on the primary coil is 20 volts, what is the output voltage?

....

....

c) A step-up transformer in the National Grid converts 11 000 volts to 220 000 volts. What is the ratio of primary turns to secondary turns?

....

Electric Circuits

1 For the following questions, put a tick in the box by the correct answer.

   a) What is the scientific unit used for energy?

      i) Watt ☐   **ii)** Joule ☐   **iii)** Volt ☐   **iv)** Ampere ☐

   b) Which unit do we use for energy used in the home?

      i) Kilowatt hour ☐   **ii)** Kilowatt ☐   **iii)** Joule ☐   **iv)** Kilo joule ☐

   c) What is the correct formula for calculating energy?

      i) $E = \dfrac{p}{t}$ ☐   **ii)** $E = \dfrac{t}{p}$ ☐   **iii)** $E = p \times t$ ☐   **iv)** $E = I \times V$ ☐

2 A 60W light bulb is switched on for 1 minute. How much energy is used? (You must include the correct units in your answer.)

....

....

3 A 3kW heater is switched on for 30 minutes. How much energy is used? (You must include the correct units in your answer.)

....

....

4 If the energy cost is 5p per kWh, how much is the cost per day (24hrs) to run a household if the average power consumption is 2.5kW?

....

5 Complete the table below.

Electrical Appliance	Energy In (J)	Useful Energy Out (J)	Efficiency %
Tumble dryer	3000	2400 (heat and kinetic)	**a)** ....
Light bulb	60	6 (light)	**b)** ....
Electric motor	400	**c)**	50%

The Wave Model of Radiation

1 Complete the following sentences.

All waves transfer _____ from one place to another without transferring _____ .

The _____ is transferred in the direction the wave _____ . There are two types of wave:

_____ and _____ .

2 a) Describe the difference between longitudinal and transverse waves.

_____

_____

_____

_____

b) Give an example of each type of wave.

i) Longitudinal: _____

ii) Transverse: _____

3 Explain what is meant by the frequency of a wave.

_____

_____

4 Use the axis below to sketch a wave. Mark up the amplitude, frequency and wavelength.

5 All electromagnetic waves travel at the same speed in a vacuum but have different frequencies. How does the wavelength depend on the frequency?

_____

_____

The Wave Model of Radiation

6 When a wave passes from one medium to another its speed will often change but its frequency stays the same. If the wave slows down, what effect does this have on the wavelength?

....

....

7 Write down the formula linking wave speed, frequency and wavelength.

....

8 For the following questions, tick the correct answer.

a) A sound wave has a wavelength of 2m and a frequency of 150Hz. What is the wave speed?

   i) 0.013m/s ☐   **ii)** 3m/s ☐   **iii)** 75m/s ☐   **iv)** 300m/s ☐   **v)** 30 000m/s ☐

b) A pebble dropped in a pond creates a wave with a frequency of 3Hz and a wavelength of 6cm. What is the wave speed?

   i) 0.18m/s ☐   **ii)** 1.8m/s ☐   **iii)** 18m/s ☐   **iv)** 0.02m/s ☐   **v)** 2m/s ☐

c) A sound wave travelling in a block of steel has a frequency of 10kHz and a wavelength of 50cm. What is the wave speed?

   i) 5m/s ☐   **ii)** 50m ☐   **iii)** 500m/s ☐   **iv)** 5000m/s ☐   **v)** 50 000m/s ☐

HT **9** Complete the following table:

Wave Speed	Frequency	Wavelength
a)	600kHz	500m
1500m/s	6kHz	**b)** ....
300 000 000m/s	**c)**	500nm (5×10^{-9}m)

10 Radio 1 transmits in the range of frequencies 97–100MHz with the exact value depending on where you live in the country.

Taking 300 000 000m/s as the wave speed of radio waves, what is the range of wavelengths that Radio 1 transmits on?

....

....

The Wave Model of Radiation

1 In which three ways can light, sound and water waves act?

a) .... **b)** .... **c)** ....

2 What happens to the speed and direction of a wavelength when a water wave crosses a boundary between one medium and another? Complete the diagram below to help you explain your answer.

....

....

....

....

Shallow water

Boundary

Deep water

3 **a)** When waves pass through a narrow gap they are diffracted. What does this mean?

....

b) What happens when a wave passes through a gap much larger than its wavelength?

....

4 The picture below shows a house in the shadow of a hill. Use ideas about diffraction to explain why the house is able to receive the long-wave radio signals from the transmitter, but light from the transmitter is unable to reach the house.

....

....

....

....

5 On the diagram below add the normal line and label the angles of incidence and reflection.

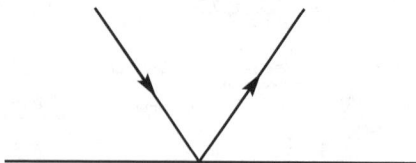

6 Explain what the **i)** incident ray, and **ii)** reflected ray is.

i) ....

ii) ....

The Wave Model of Radiation

1 On the diagrams below, sketch the path of a light ray as it enters and leaves the glass block. Label the normal and the angles of incidence and refraction.

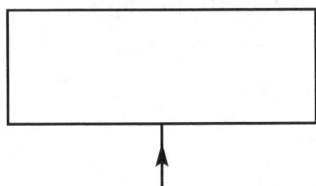

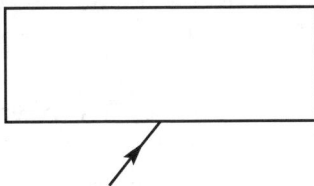

2 Explain why a light wave changes direction when it passes from one medium into another.

_____

_____

_____

3 When light travels from a dense medium to a less dense medium, e.g. glass into air, the angle of refraction can be greater than 90°. What is this called and what happens to the light ray?

_____

_____

4 Are the following statements **true** or **false**?

a) When waves are in step their amplitudes add up. _____

b) When waves are out of step their amplitudes add up. _____

c) Destructive interference means that the waves explode. _____

d) Constructive interference means that the amplitudes of the waves add up. _____

5 Freak waves can sometimes occur when waves at sea constructively interfere. Why can this be dangerous for ships?

_____

_____

_____

The Wave Model of Radiation

1 What is a photon?

_____

2 Number the following wavelengths 1–7 to put them into the correct order, starting with the highest frequency and photon energy first.

a) Microwaves ☐　　**b)** X-rays ☐　　**c)** Radio waves ☐　　**d)** Ultraviolet ☐

e) Gamma rays ☐　　**f)** Visible light ☐　　**g)** Infrared rays ☐

3 What are the three main differences between sound waves and electromagnetic waves?

a) _____

_____

b) _____

_____

c) _____

_____

4 At what speed do electromagnetic waves travel through space?

_____

HT

5 The power of a beam of radiation depends on two things. One is the number of photons delivered by the beam of radiation every second. What is the other?

_____

_____

6 What two facts support the idea that electromagnetic radiation travel as waves?

a) _____

b) _____

The Wave Model of Radiation

7 a) On the diagram below, show what happens to white light as it passes through the prism.

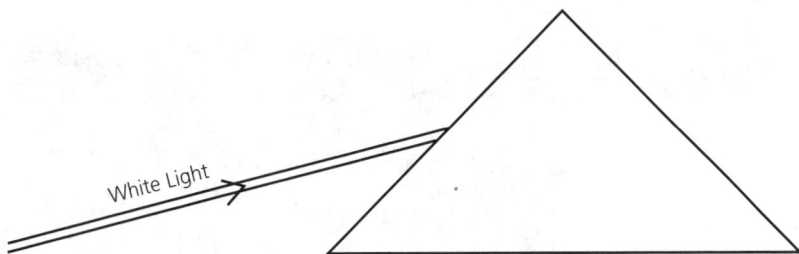

White Light →

b) Explain your answer to part a).

_____

_____

8 Match the types of radiation to their uses.

Type of Radiation
Radio Waves
Light and Infrared
Microwaves
X-rays

Use
Taking shadow pictures of bones.
Transmitting radio and TV programmes.
Satellite communications. Heating food.
Carrying information on computer networks and telephone cables.

9 Explain how the properties of microwaves make them suitable for sending signals through the Earth's atmosphere.

_____

_____

_____

10 Explain how the properties of X-rays make them suitable for their use(s).

_____

_____

_____

The Wave Model of Radiation

1 For a wave to carry a signal it must be modulated. Complete the table below with a description and a diagram showing the output wave of the two types of modulation.

Description	Diagram
a)	
Frequency Modulation. The frequency of the output wave changes but the amplitude remains constant.	**b)**

2 Explain the difference between an analogue and a digital signal. Draw diagrams to help you explain your answer.

....

....

....

3 On the axis below, sketch the digital signal represented by 1011001.

The Wave Model of Radiation

4 On the axis below sketch an analogue wave with a wavelength of 2cm and an amplitude of 1cm.

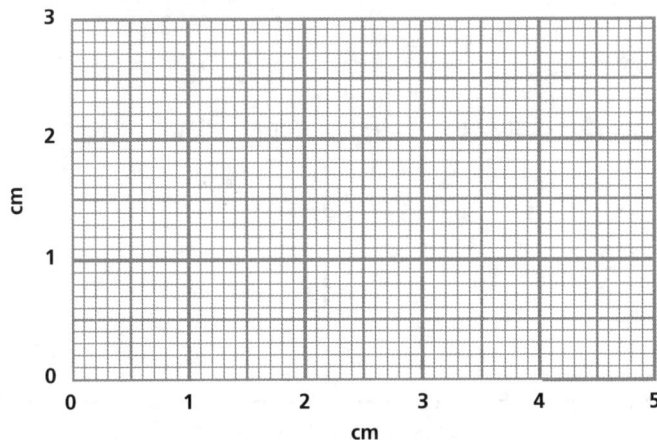

5 When talking about signals we often talk about 'noise'. What do we mean by **signal noise**?

....

....

6 The two diagrams below show analogue and digital signals which have been distorted by noise. Explain why digital signals are more tolerant of noise and why this is a benefit.

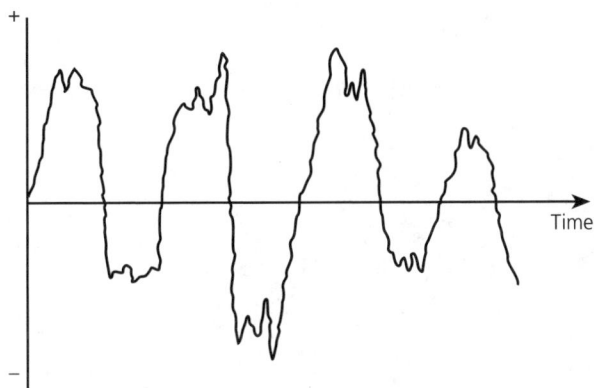

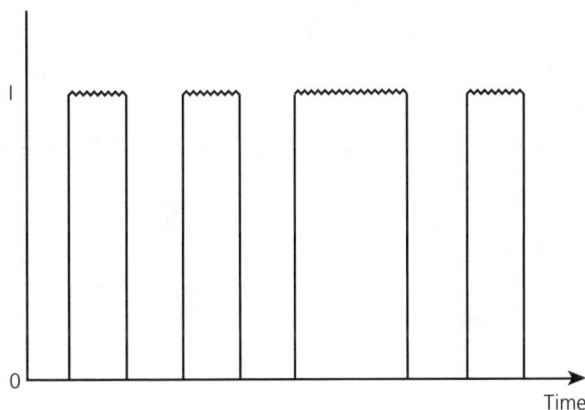

....

....

....

....

Further Physics, Observing the Universe

1 In which direction do the stars *appear* to move?

East to west

2 *The Sun moves across the sky from east–west*. Is this statement **true** or **false**? Explain your answer.

False. The sun moves from west to east. It is due to the rotation of the earth that the sun appears to move from east-west.

3 Complete the following sentence.

Stars appear to travel ___east-west___ across the sky in ___23___ hours and

___6___ minutes.

4 Explain the difference between a sidereal day and a solar day.

A sidereal day is the time it takes for the earth to rotate 360° on it's axis. A solar day is the time it takes for the earth to rotate in 24 hours.

5 Briefly explain why a sidereal day is shorter than a solar day.

A sidereal day is 4 minutes shorter than a solar day because the earth needs to orbit the sun and needs to rotate a bit more to be directly overhead the sun.

6 How much shorter is a sidereal day compared with a solar day? Tick the correct option.

a) 2 minutes ☐

b) 3 minutes ☐

c) 4 minutes ☑

d) 5 minutes ☐

Further Physics, Observing the Universe

1 Explain, in as much detail as you can, why some stars are only visible at certain times of the year.

> some stars are only visible at certain times of year
> because of the movement of the earth around the sun

2 Fill in the gaps to complete the following sentences.

Stars with a _positive_ declination are visible from the northern hemisphere.

Stars with a _negative_ declination are visible from the southern hemisphere.

If a star has a zero declination, it can be seen above the _equator_.

3 As well as a star's declination, what other measurement can be used to describe its position?

> ascension

4 *Mercury, Venus and Neptune are all stars that can be seen with the naked eye.* Is this statement **true** or **false**? Explain your answer.

> True because the planets look similar to stars but they
> change their positions in complicated patterns when compared to
> the background of fixed stars

5 Explain why the apparent movement of planets across the sky is different to that of the stars.

HT **6** A planet's apparent movement is east–west when it is on the same side of the Sun as the Earth. What will be the planet's apparent direction of movement when it is on the opposite side of the Sun to the Earth? Explain why this is so.

Further Physics, Observing the Universe

1 How long does it take for the Moon to apparently travel east–west once across the sky? Tick the correct option.

a) Just under 24 hours ☐

b) Exactly 24 hours ☐

c) Just over 24 hours ☑

d) 28 days ☐

HT

2 Complete the following sentence.

The Moon appears to travel east–west across the sky in _____24_____ hours and

_____49_____ minutes.

3 Are the following statements **true** or **false**?

a) During a new Moon, the Moon appears dark because it is in the Earth's shadow. _____True_____

b) The Moon changes shape during the lunar cycle. _____False_____

c) Half of the Moon is always in darkness. _____False_____

d) A full Moon occurs when the Moon is on the opposite side of the Earth to the Sun. _____True_____

4 Explain why the Moon's appearance changes as it orbits the Earth.

Because we cannot see all the sides that is lit by the sun

5 Explain the difference between a partial solar eclipse and a total solar eclipse.

A partial solar eclipse is when the moon passes between the
earth and sun. A total solar eclipse occurs when the moon is
directly infront of the sun.

Further Physics, Observing the Universe

6 Sketch a ray diagram to illustrate the formation of a total solar eclipse.

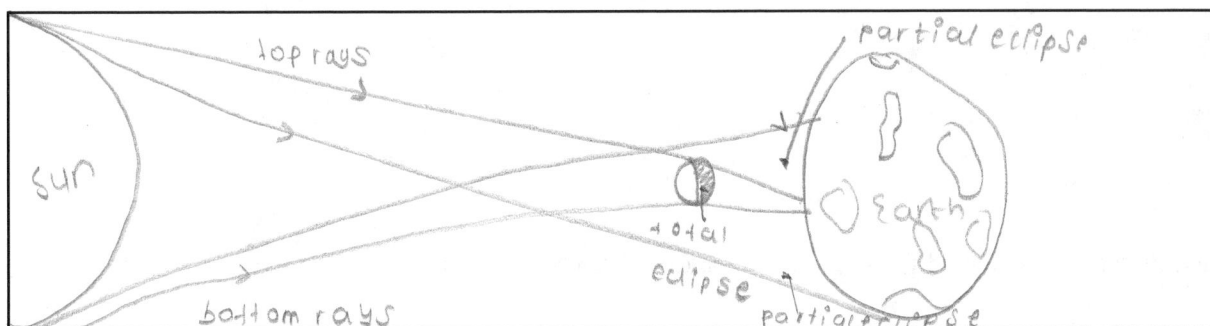

7 Explain when a lunar eclipse would occur.

When the earth is between the sun and the moon

8 Explain why total lunar eclipses are seen much more often than total solar eclipses.

9 Explain why a solar eclipse will never occur during a half Moon.

HT

10 What is the **ecliptic**?

The apparent path the sun traces out along the sky

11 a) How often approximately do solar eclipses occur?

2-5 every year

b) Explain why solar eclipses do not occur very often. Draw a diagram to help explain your answer.

Because the moon does not orbit the earth in the same plane as the earth orbits the sun

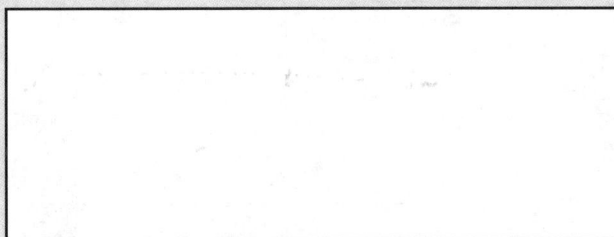

Further Physics, Observing the Universe

1 What is another name for a convex lens?

converging lens

2 What is the focal point of a lens?

the point where light entering will be bought to a focus

3 In which direction are rays of light passing through a convex lens bent?

inwards

4 Use the words below to complete the following sentences about convex lenses.

pronounced   **short**   **long**   **slight**

A weak lens has a _____*slight*_____ curvature and a _____*longer*_____ focal length.

A strong lens has a _____*pronounced*_____ curvature and a _____*short*_____ focal length.

5 What unit is used to measure the power of a lens?

dioptres

6 If a convex lens has a focal length of 50cm, what will be the power of the lens? Show your workings.

power= 1/0.5   5

Power = 2 dioptres

7 Calculate the power of a lens if the focal length is 20m. Show your workings.

Power= 1/20

power = 0.05 dioptres

8 Complete the ray diagram below to show the formation of an image when the object is located at 2F.

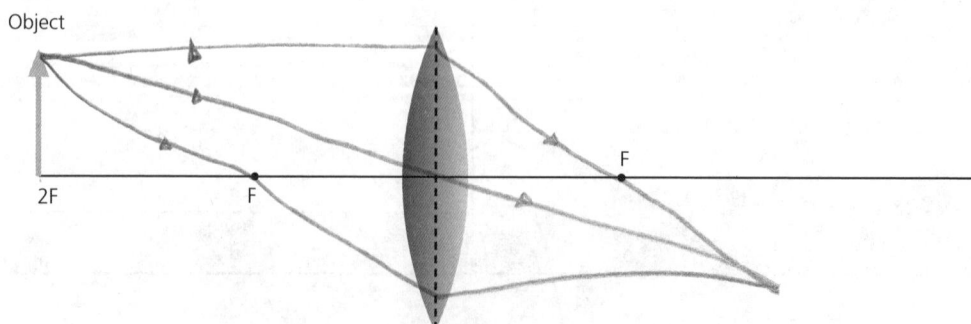

Further Physics, Observing the Universe

1 a) What are the names of the two lenses used in a simple telescope?

i) _eye piece_   **ii)** _objective_

b) Which of the lenses given in your answer to part a) is the most curved?

_____

2 Sketch a diagram illustrating the path of the light rays through a simple refracting telescope. Label the two different lenses.

3 Some telescopes use a concave mirror instead of a convex lens. What is the advantage of using a mirror in terms of the images that can be seen?

It allows the image to be larger

4 On the diagram of the reflecting telescope, label the objective lens(es) and the eyepiece lens.

objective lens

eyepiece

HT

5 How does angular magnification make an object appear?

makes an image bigger

6 A telescope has an objective lens with a focal length of 10m, and an eyepiece lens with a focal length of 50cm. Calculate the magnification of the telescope.

Magnification: ${}^{10}/_{0.5} = 20$

7 A simple telescope has an eyepiece lens with a power of 100 dioptres and an objective lens with a power of 2 dioptres. Calculate the...

a) focal length of the lenses _$100 = {}^{1}/_{FL} = 0.01$_

b) magnification of the telescope _$2 = {}^{1}/_{FL} = 0.5$ $M = {}^{0.5}/_{0.01} = 50$_

Further Physics, Observing the Universe

1 Parallax can be *thought of* as the apparent motion of an object against a background. What actually causes the parallax motion of an object?

motion of the observer

2 If you hold your hand out in front of you, with your thumb sticking up, and close one eye then the other eye, you thumb appears to move. What is actually happening?

Your looking at it from a different angle

3 Give a definition for **parallax angle**.

half the angle ~~moveme~~ moved against a background of

distant stars in 6 months

4 Are the following statements about parallax **true** or **false**?

a) Parallax can only be used to measure the distance to planets. False

b) An object that is further away from Earth will have a larger parallax angle than a closer object. False

c) The observations used to find the parallax angle are made on opposite sides of the Earth. True

5 a) On the diagram below, draw and label the parallax angle.

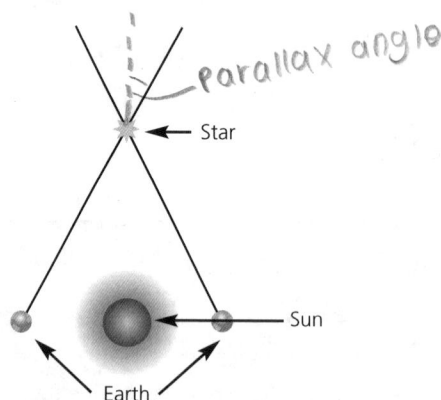

Parallax angle

Star

Sun

Earth

b) What would happen to the parallax angle if the star was further away from Earth?

smaller angle

c) When using parallax to measure the distance to planets in our solar system, astronomers have to arrange two simultaneous observations from opposite sides of the Earth. Why is this important?

© Lonsda

Further Physics, Observing the Universe

1 Give the formula used to calculate interstellar distances. (Make sure you include the correct units.)

Distance (parsecs) = 1/parallax angle (arcsecond)

2 A distant star has an observed parallax angle of 0.5 arcseconds. How far away is the star? Show your workings.

D = 1/0.5

D = 2 parsecs

3 An astronomer observes a star that is 5 parsecs away. Calculate the observed parallax angle. Show your workings.

5 = 1/pa

pa = 0.2 arcseconds

4 Explain why it is only possible to use parallax to measure the distance to relatively close stars and not distant galaxies.

Because the parallax angle is too small to measure. Astronomers need to use mega parsecs

5 An astronomer on Earth measures the parallax angle of two different stars. Star A has a parallax angle of 0.7 arcseconds and star B has a parallax angle of 0.3 arcseconds.

a) Which star is furthest away from Earth?

Star A

b) If the stars are seen on opposite sides of our solar system, how far apart are they?

6 Complete the table below.

Star	Distance (pc)	Parallax Angle (arcseconds)
Sirius	2.64	**a)**
Arcturus	**b)**	0.044
Barnard's Star	**c)**	0.55
Proxima Centauri	1.30	**d)**

Further Physics, Observing the Universe

7 *The parallax angle is half the angle moved against the background of distant stars in one year.* Is this statement

true or **false**? _False_

8 For what kinds of objects would the distance be measured in Mpc?

bigger intergalactic distances

9 *A closer star will appear brighter than a more distant star.* When is this statement correct? (Circle) the correct option.

~~**Always**~~   **Sometimes**   **Never**

10 Complete the following paragraph.

All stars have an _intrinsic_ brightness that is a measure of how much _energy_

they are emitting. This brightness depends on each star's _size_ and _temperature_ .

11 a) What two things does the observed brightness of a star depend on?

i) _intrinsic brightness_   **ii)** _Distance from earth_

b) What makes Cepheid variable stars different from other stars?

Does not have a constant intrinsic brightness

c) What does the frequency of a Cepheid variable star allow us to estimate?

intrinsic brightness

d) Complete the table showing the different methods of measuring astronomical distances.

Method	Advantage	Disadvantage
Parallax	Accurate method of measurement	**i)**
Observing Brightness	Very simple method. The brighter a star, the further away it is.	**ii)**
Using Cepheid variables	**iii)**	Not all stars are Cepheid variables so this method only works for a few stars

Further Physics, Observing the Universe

1 a) In 1920, a famous debate about the scale of the Universe took place. Give the names of the two main astronomers involved in the debate.

i) _curtis_

ii) _shapely_

b) What was the name originally given to the 'fuzzy objects' observed in the night sky that played a major role in the debate?

nebulae

c) Outline the two opposing opinions about what these fuzzy objects were.

curtis believed it was distant galaxies

shapley believed it was the milky way

2 Fill in the gaps to complete the paragraph below

The debate was finally decided when Edwin Hubble observed _cepheid_ variable stars in one of the fuzzy objects. His observations showed that the stars were _____ distant than any stars in the _milky_ _way_ galaxy. This, therefore, supported the idea that the objects must be different _galaxies_.

3 Give the formula linking the Hubble constant, speed of recession and distance. (Give both units of measure.)

4 A galaxy is a distance of 80Mpc from Earth. If the hubble constant is 70 kms^{-1} Mpc^{-1}, calculate the speed of recession.

HT

5 Observations of a distant galaxy give a speed of recession of 14 000km/s. If the Hubble constant is 70 kms^{-1} Mpc^{-1}, calculate the distance to the galaxy in Mpc.

6 An observed galaxy has a speed of recession of 46 000km/s. If the Hubble constant is 2.3 x 10^{-18} s^{-1}, calculate the distance to the galaxy in km.

Further Physics, Observing the Universe

1 *The term* **fluid** *is only used to describe a liquid*. Is this statement **true** or **false**? Explain your answer.

....

2 Complete the following paragraph.

When .... in a fluid collide with something they exert a .... which

is felt as ....

3 What will happen to the fluid pressure if the particles collide at a higher speed?

....

4 a) What is meant by the term **absolute zero**?

when ....

b) What unit is used to measure absolute temperature?

....

5 The absolute temperature of a gas is doubled. If the volume remains constant, what effect will this have on the pressure?

....

6 Are the following statements **true** or **false**?

a) Heavy particles exert more pressure than lighter particles. ....

b) Faster particles exert more pressure than slower particles. ....

c) Increasing the volume will increase the pressure. ....

d) Increasing the temperature will increase the pressure. ....

7 Complete the table below.

Temperature (Kelvin)	Temperature (degrees Celsius)
0	**a)**
273	**b)**
c)	37
d)	100

Further Physics, Observing the Universe

1 a) Label the diagram of a star.

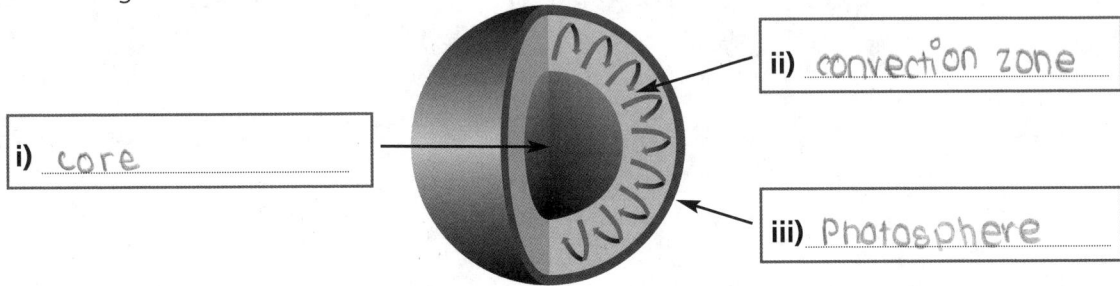

ii) convection zone

i) core

iii) Photosphere

b) Which part of the star is the hottest? core

c) Which part of the star radiates energy into space? Photosphere

d) In what part of a star does fusion take place?

2 Are the following statements **true** or **false**?

a) The Sun is a star. True

b) Stars emit electromagnetic radiation.

c) The temperature of a star affects the frequency of radiation it emits.

3 Why does a star produce a continuous emission spectrum?

4 Using the words below to help you, complete the following sentences.

spectra   **absorbed**   **elements**   **lines**

All stars produce a continuous emission spectrum except for a few blank .... which

occur because these parts of the spectrum are .... by the material in the star. We can

determine the composition of a star by comparing the star's spectrum with .... from

various ....

5 How would the spectrum of a star that contains several different elements be different from one that
contains just hydrogen?

Further Physics, Observing the Universe

1. Explain, in as much detail as you can, how a star is formed.

   ....

   ....

   ....

   ....

2. a) What process happens to a protostar in order for it to become a main sequence star?

   ....

   ....

   b) What is released from the star during this process?

   ....

3. What happens towards the end of a star's life in terms of its 'fuel'?

   ....

4. What happens to a red giant after all its helium has been used up?

   ....

5. What is a supernova?

   ....

6. A very large star is close to the end of its life. What two things could it become?

   a) ....

   b) ....

7. Why are average sized stars unable to form heavy nuclei?

   ....

   ....

   ....

Further Physics, Observing the Universe

1 a) In 1911, a groundbreaking experiment provided evidence about the structure of the atom.

   i) Which three scientists carried out the experiment?

   ....

   ii) What metal was used for the target in the experiment? ....

   iii) What type of particles were used to bombard the target? ....

b) Draw lines to link each observation with its conclusion.

Observation	Conclusion
Most of the alpha particles pass straight through.	There is a heavy positive nucleus that takes up a very small amount of space.
Some of the alpha particles are deflected.	Most of the gold is empty space.
A very small percentage of the alpha particles are reflected.	There are charges concentrated in a few places within the gold.

c) Draw lines to link each conclusion with its explanation.

Conclusion	Explanation
There is a heavy positive nucleus that takes up a very small amount of space.	Alpha particles are relatively heavy and would knock light particles out of the way. They are positive and like charges repel.
Most of the gold is empty space.	Charges are repelled or attracted if they pass close to other charges.
There are charges concentrated in a few places within the gold.	All alpha particles would be reflected or deflected if this was not true.

2 What force holds the nucleus together? ....

3 Why does it require a large amount of energy to force two protons together?

....

....

4 Are the following statements **true** or **false**?

a) The nuclear force is stronger than the electrostatic force. ....

b) The nuclear force has a longer range than the electrostatic force. ....

Further Physics, Observing the Universe

1 Name four types of telescopes used by astronomers.

a) ....

b) ....

c) ....

d) ....

2 Explain why most ground-based telescopes are situated at high altitude.

....

....

3 Why are ground-based telescopes not situated close to major cities?

....

....

....

4 As well as image quality, give two other factors that should be considered when deciding on the location for a ground-based telescope.

a) ....

b) ....

5 For each of the following situations, circle the most appropriate type of telescope that would be used.

a) A well-funded project that wants to get the best possible visual images.

    Space-based   **Ground-based radio telescope**   **Ground-based optical**

b) For looking at dead stars that are too cool to give off visible light.

    Space-based   **Ground-based radio telescope**   **Ground-based optical**

c) To look into the deepest regions of space.

    Space-based   **Ground-based radio telescope**   **Ground-based optical**

d) To produce visible images, with low maintenance costs.

    Space-based   **Ground-based radio telescope**   **Ground-based optical**

Further Physics, Observing the Universe

6 a) Give two advantages of using space-based telescopes.

   i) ....

   ii) ....

b) Give two disadvantages of using space-based telescopes.

   i) ....

   ii) ....

7 Many modern telescopes are robotic and can be controlled from a remote location. Give three advantages of using this type of telescope.

a) ....

b) ....

c) ....

HT

8 Sketch a diagram showing diffraction of waves passing through a narrow gap.

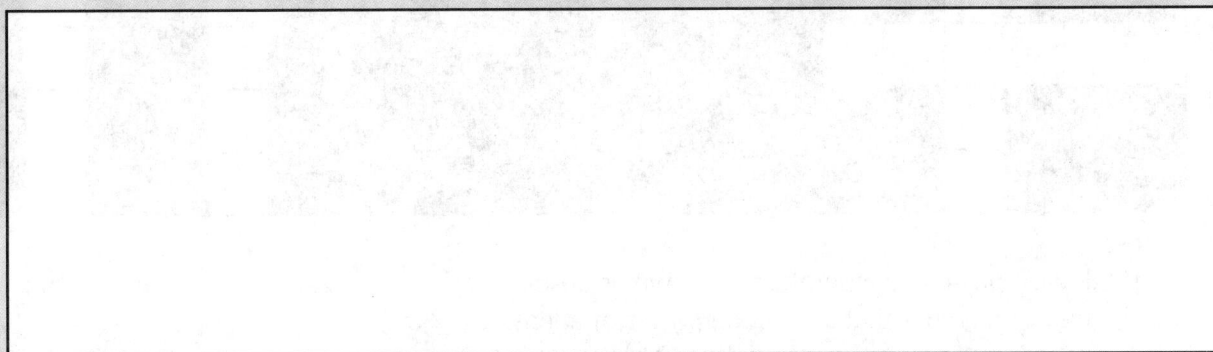

9 What consequence does diffraction have on images formed by radio telescopes?

....

....

10 Why are optical telescopes able to produce very sharp images?

....

....

Further Physics, Observing the Universe

Across

5. The lowest possible temperature is known as absolute _____ . (4)
6. The unit of measure that shows the power of a lens. (8)
8. A _____ day is a full 24 hours. (5)
9. An imaginary line that connects the Earth to the Sun. (8)
11. A star with a brightness that varies with a fixed frequency. (7)
14. The declination of a star visible from the southern hemisphere. (8)
15. The angle that describes the positions of the stars relative to a fixed point on the equator. (11)
16. The process that releases energy in the Sun. (6)

Down

1. The charge that a nucleus has. (8)
2. A _____ day is the time it takes the Earth to rotate 360°. (8)
3. An exploding star. (9)
4. A more curved lens is _____ powerful than a flatter lens. (4)
7. This causes waves to spread out as they pass through an aperture. (11)
8. Telescopes placed here do not suffer from atmospheric interference. (5)
10. The inner part of a star. (4)
12. The apparent movement of a near object against a distant background.(8)
13. Heating a gas makes the pressure _____ . (8)

1 The information below is about using electromagnetic radiation and chemical analysis to detect explosives. Read the information and answer the questions that follow.

Following the recent foiling of an alleged terrorist plot to blow up transatlantic passenger flights using the liquid explosive triacetone triperoxide (TATP), attention is being focused on new detection and scanning technologies.

The fear is that explosives could be disguised in everyday objects such as fizzy-drink bottles. However, explosives experts say that extra security measures were put in place as an extra precaution, not because liquid explosives are difficult to detect. In fact, liquid explosives could be easier to detect as they have to be kept in a container, whereas solid explosives come in a variety of shapes and sizes and have no defined form.

Chemical Analysis

Although X-ray machines are used to search luggage, explosive equipment such as detonators can be hidden inside electronic equipment, so chemical analysis can be used to detect them. A swab can be taken from a bag and placed into a machine that heats up the sample and performs a spectrographic analysis of the vapours. The machine searches for traces of nitrogen, as this is found in the majority of explosives.

Experts say that although it would not be possible to perform a careful chemical analysis on every bottle of liquid to be carried onto a plane, it might be possible to ban any non water-based liquids and have a simple test to pick these out using universal indicator paper.

Electronic Techniques

Experts say improved airport scanning that detects explosives remains a top priority. The most commonly used X-ray scanning devices are used to detect suspicious shapes, such as the pattern of wires likely to be found in a bomb. But there may be the need for even more sophisticated scanning techniques. Some new machines can detect specific compounds by measuring reflected X-ray photons. They can reveal materials made up of elements that have a low atomic number (as explosives have) such as hydrogen and nitrogen.

Another more sophisticated detection method is scanning with terahertz waves, which lie between microwaves and infrared on the electromagnetic spectrum. At the moment these scanners are very large and their use in airports would be impractical, but portable scanners are being developed. These scanners could measure the way objects absorb and reflect terahertz waves, as they pass straight through plastic, fabric, wood and stone but can be used to spot other compounds, including certain drugs, metals and explosives.

Ideas in Context

1 **(continued)**

a) What do X-ray scanners do?

....

b) How do X-ray scanners work? Place a tick beside the correct answer.

i) X-rays are absorbed by dense materials. ☐

ii) X-rays are emitted by dense materials. ☐

iii) X-rays are reflected by dense materials. ☐

c) Would a liquid explosive be harder to detect than a solid explosive by an X-ray machine? Explain your answer.

....

2 What could chemical analysis be used for?

....

....

....

....

3 **a)** Name two elements commonly used in explosives.

i) ....

ii) ....

b) What do these elements have in common?

....

4 What is a photon?

....

5 If terahertz radiation is between infrared and microwaves in the electromagnetic spectrum, does it have a low or high frequency?

....

....

© Lonsda

Across

1. A point when the Earth's shadow is cast on the Moon. (5, 7)
3. The measurement used for interstellar distances. (6)
6. A _____ variable star has a changing variable brightness. (7)
8. The point at which all rays of light meet. (5)
11. The change in direction and speed of a wave as it passes from one medium into another. (10)
12. A positively or negatively charged particle. (3)
13. The fundamental unit of a living organism. (4)
14. A circuit in which there is only one path for the current to take. (6)
15. The layer of gas surrounding the Earth. (10)

Down

2. When an oceanic plate is forced under a continental plate. (10)
3. A packet of energy. (6)
4. A wave that carries a signal. (7)
5. The lowest temperature possible (in Kelvins). (4)
7. A _____ day is the time it takes the Earth to rotate 360°. (8)
8. A push or pull acting on an object. (5)
9. A signal that has two fixed values. (6)
10. A collection of galaxies. (8)

Notes

Acknowledgements

Acknowledgements

The author and publisher would like to thank everyone who has contributed to this book:

IFC   ©iStockphoto.com / Andrei Tchernov

ISBN 978-1-906415-00-6

Published by Letts and Lonsdale.

Author: Nathan Goodman

Project Editor: Charlotte Christensen

Cover and Concept Design: Sarah Duxbury

Designer: Joanne Hatfield

Letts and Lonsdale make every effort to ensure that all paper used in our books is made from wood pulp obtained from sustainable and well-managed forests.

Author Information

Nathan Goodman has an in-depth understanding of the new science specifications, thanks to his roles as Secondary Science Strategy Consultant for North East Lincolnshire LEA and Regional Coordinator at the Institute of Physics for the physics teacher network. As Assistant Headteacher, Nathan is involved in improving the teaching and learning of science at his current school.